Robust Formation Control for Multiple Unmanned Aerial Vehicles

This book is based on the authors' recent research results on formation control problems, including time-varying formation, communication delays, and fault-tolerant formation, for multiple UAV systems with high nonlinearities, parameter uncertainties, and external disturbances.

Differentiating from existing works, this book presents a robust optimal formation approach to designing distributed cooperative control laws for a group of UAVs based on the linear quadratic regulator control method and the robust compensation theory. The proposed control method is composed of two parts: the nominal part to achieve desired tracking performance and the robust compensation part to restrain the influence of high nonlinearities, parameter uncertainties, and external disturbances on the global closed-loop control system. Furthermore, this book gives proof of their robust properties. The influence of communication delays and actuator fault tolerance can be restrained by the proposed robust formation control protocol, and the formation tracking errors can converge into a neighborhood of the origin bounded by a given constant in a finite time. Moreover, this book provides details about the practical application of the proposed method to design formation control systems for multiple quadrotors and tail-sitters. Additional features include a robust control method that is proposed to address the formation control problem for UAVs, and theoretical and experimental research for the cooperative flight of the quadrotor UAV group and the tail-sitter UAV group.

Robust Formation Control for Multiple Unmanned Aerial Vehicles is suitable for graduate students, researchers, and engineers in the system and control community, especially those engaged in the areas of robust control, UAV swarming, and multi-agent systems.

Automation and Control Engineering

Series Editors:
Frank L. Lewis, Shuzhi Sam Ge, and Stjepan Bogdan

Adaptive and Fault-Tolerant Control of Underactuated Nonlinear Systems
Jiangshuai Huang and Yong-Duan Song

Discrete-Time Recurrent Neural Control
Analysis and Application
Edgar N. Sánchez

Control of Nonlinear Systems via PI, PD and PID
Stability and Performance
Yong-Duan Song

Multi-Agent Systems
Platoon Control and Non-Fragile Quantized Consensus
Xiang-Gui Guo, Jian-Liang Wang, Fang Liao, and Rodney Swee Huat Teo

Classical Feedback Control with Nonlinear Multi-Loop Systems
With MATLAB® and Simulink®, Third Edition
Boris J. Lurie and Paul Enright

Motion Control of Functionally Related Systems
Tarik Uzunović and Asif Sabanović

Intelligent Fault Diagnosis and Accommodation Control
Sunan Huang, Kok Kiong Tan, Poi Voon Er, and Tong Heng Lee

Nonlinear Pinning Control of Complex Dynamical Networks
Edgar N. Sanchez, Carlos J. Vega, Oscar J. Suarez, and Guanrong Chen

Adaptive Control of Dynamic Systems with Uncertainty and Quantization
Jing Zhou, Lantao Xing, and Changyun Wen

Robust Formation Control for Multiple Unmanned Aerial Vehicles
Hao Liu, Deyuan Liu, Yan Wan, Kimon P. Valvanis, and Frank L. Lewis

For more information about this series, please visit: https://www.crcpress.com/Automation-and-Control-Engineering/book-series/CRCAUTCONENG

Robust Formation Control for Multiple Unmanned Aerial Vehicles

Hao Liu, Deyuan Liu, Yan Wan,
Kimon P. Valavanis, and Frank L. Lewis

CRC Press is an imprint of the
Taylor & Francis Group, an **informa** business

First edition published 2023
by CRC Press
6000 Broken Sound Parkway NW, Suite 300, Boca Raton, FL 33487-2742

and by CRC Press
4 Park Square, Milton Park, Abingdon, Oxon, OX14 4RN

CRC Press is an imprint of Taylor & Francis Group, LLC

ISBN: 9781032149400 (hbk)
ISBN: 9781032150246 (pbk)
ISBN: 9781003242147 (ebk)

DOI: 10.1201/9781003242147

Typeset in Palatino
by codeMantra

Contents

Preface

Unmanned Aerial Vehicles (UAVs) have various applications, such as surveillance, border patrol, building exploration, mapping, and inspection, where the operating environments can be dangerous and inaccessible. In practical applications, it is difficult for a single UAV to perform complex tasks, due to the limited endurance, load capacity, and coverage radius. Compared to a single aerial vehicle, multiple UAVs are capable to perform complex tasks, simultaneously, which can increase efficiency and provide redundancy against individual failures. Therefore, formation flight of a team of UAVs has attracted considerable attention in both the military and civil fields over the past two decades. It is significant for achieving the desired formation flying to design robust formation controllers.

There are multiple challenges for the robust formation controller design of a group of UAVs. The dynamics of UAVs is highly nonlinear and coupled. Multiple uncertainties such as unmodelled uncertainties, parametric perturbations, and external disturbances can affect the tracking performance of the flight control system. Furthermore, in practical applications, such as source seeking and target enclosing, the UAV formation is required to track the desired trajectory to perform the tasks while keeping the desired time-varying formation. Moreover, individual UAVs cooperate with others through communication for information sharing. During the information exchange, time delays inevitably exist, which may degrade the control performance of the formation and even destabilize the entire system. In addition, each UAV may be subject to certain actuator failures. In the framework of formation control, the actuator failure in an individual UAV can spread over neighboring UAVs through the interaction topology and affect the performance of the whole system. All of the above issues make it difficult to achieve desired formation control performance for a group of UAVs.

This book is based on the authors' recent research results on formation control problems, including time-varying formation, communication delays, and fault-tolerant formation, for multiple UAV systems with high nonlinearities, parameter uncertainties, and external disturbances. Different from the existing works, we present a robust optimal formation approach to design distributed cooperative control laws for a group of UAVs, based on the linear quadratic regulator (LQR) control method and the robust compensation theory. The proposed control method is composed of two parts: the nominal part to achieve desired tracking performance and the robust compensation part to restrain the influence of high nonlinearities, parameter uncertainties, and external disturbances on the global closed-loop control system. Furthermore, this book gives the proof of robust properties. The influence of communication delays and actuator fault tolerance can be restrained by the proposed robust formation control protocol, and the formation tracking errors can converge into a neighborhood of the origin bounded by a given constant in a finite time. Moreover, this book provides details about the practical applications of the proposed method to design formation control systems for multiple quadrotors and tail-sitters.

The main features of this book include a robust control method proposed to address the formation control problem for UAVs, and theoretical and experimental research for the cooperative flight of the quadrotor UAV group and the tail-sitter UAV group. This book is suitable for graduate students, researchers, and engineers in the system and control community, especially those engaged in the area of robust control, UAV swarming, and multi-agent systems.

We are grateful to Beihang University for providing resources for our research work. We also gratefully acknowledge the support of our research by the National Natural Science Foundation of China (grant nos. 61873012, 61503012, and 61633007), China Postdoctoral Science Foundation (grant nos. 2021M700336), and National Science Foundation (grant nos. 1730675 and 1714519). Finally, we would like to thank the entire team at Taylor & Francis Group for their cooperation and encouragement in bringing out the work in the form of a monograph.

Authors

Hao Liu received a B.E. degree in control science and engineering from the Northwestern Polytechnical University, Xi'an, China, in 2008, and the Ph.D. degree in automatic control from Tsinghua University, Beijing, China, in 2013. In 2012, he was a visiting student in the Research School of Engineering, Australian National University. From 2013 to 2020, he has been with the School of Astronautics, Beihang University, Beijing, China, where he is currently an associate professor. Since 2020, he has been with the Institute of Artificial Intelligence, Beihang University, Beijing, China. From 2017 to 2018, he was a visiting scholar at the University of Texas at Arlington Research Institute, Fort Worth, USA. He received the best paper award in IEEE ICCA 2018. His research interests include formation control, reinforcement learning, robust control, nonlinear control, unmanned aerial vehicles, unmanned underwater vehicles, and multi-agent systems. He serves as an associate editor of *Transactions of the Institute of Measurement and Control* and *Advanced Control for Applications: Engineering and Industrial Systems.*

Deyuan Liu received a B.E. degree in automation from the Beijing University of Chemical Technology, Beijing, China, in 2015, and a Ph.D. degree in flight vehicle design from the School of Astronautics, Beihang University, Beijing, China, in 2021. He is currently a postdoctoral fellow of the Zhuoyue Program in control theory and control engineering with Beihang University, Beijing, China. His current research interests include multi-agent systems, robust control, nonlinear control, formation control, and tail-sitter aircraft control.

Yan Wan is currently a distinguished university professor in the Electrical Engineering Department at the University of Texas at Arlington. She received her Ph.D. degree in electrical engineering from Washington State University in 2009 and then did postdoctoral training at the University of California, Santa Barbara. Her research interests lie in the modeling, evaluation, and control of large-scale dynamical networks, cyber-physical systems, stochastic networks, and their applications to smart grids, urban aerial mobility, autonomous driving, robot networking, and air traffic management. She is an appointed member of the Board of Governors of the IEEE Control Systems Society (CSS) and serves on the Conference Editorial Board and Technology Conference Editorial Board. She is also a technical committee member of AIAA Intelligent Systems, IEEE CSS Nonlinear Systems and Control, and IEEE CSS Networks and Communication Systems.

Kimon P. Valavanis received a Diploma degree in electrical and electronic engineering from the National Technical University of Athens, Athens, Greece, in 1981, a M.Sc. degree in electrical engineering, and a Ph.D. degree in computer and systems engineering from Rensselaer Polytechnic Institute, Troy, NY, USA, in 1984 and 1986, respectively. He is currently a professor and chair of the Electrical and Computer Engineering Department and also the acting chair of the Computer Science Department, University of Denver, Denver, CO, USA. His current research interests include unmanned systems and distributed intelligence systems.

Frank L. Lewis is a member of the National Academy of Inventors, a fellow of IEEE/IFAC/UK/Institute of Measurement & Control, a PE Texas, and a UK Chartered Engineer. He is a UTA Distinguished Scholar Professor, UTA Distinguished Teaching Professor, and Moncrief-O'Donnell Chair at the University of Texas at Arlington Research Institute. He received a bachelor's degree in physics/EE in 1971, an M.S.E.E. degree in 1971 from Rice University, Houston, TX, USA, and an M.S. degree in aeronautical engineering in 1977 from the University of West Florida, Pensacola, FL, USA, and a Ph.D. degree in electrical engineering in 1981 from the Georgia Institute of Technology, Atlanta, GA, USA. He works in feedback control, intelligent systems, cooperative control systems, and nonlinear systems. He is the author of 7 U.S. patents, numerous journal special issues, journal papers, and 20 books, including *Optimal Control, Aircraft Control, Optimal Estimation,* and *Robot Manipulator Control* which are used as university textbooks worldwide. He received the Fulbright Research Award, NSF Research Initiation Grant, ASEE Terman Award, Int. Neural Network Soc. Gabor Award, U.K. Inst Measurement & Control Honeywell Field Engineering Medal, IEEE Computational Intelligence Society Neural Networks Pioneer Award, and AIAA Intelligent Systems Award. He also received Outstanding Service Award from Dallas IEEE Section and Texas Regents Outstanding Teaching Award 2013, and was selected as Engineer of the Year by Ft. Worth IEEE Section. He was listed in Ft. Worth Business Press Top 200 Leaders in Manufacturing.

1

Introduction and Background

1.1 Background

In the past decades, multi-agent cooperative control has become the frontier of artificial intelligence. The objective of multi-agent control is to divide the complex system into several small individuals, which can exchange information and works cooperatively. As a typical agent platform, an unmanned aerial vehicle (UAV) is widely used in the research of multi-agent cooperative control. In particular, the formation control problem for a group of UAV systems has attracted significant attention recently in multiple fields, such as robotics, aerospace, and wireless communication. UAV formation has been widely used in multiple missions, such as persistent surveillance, drag reduction, exploration, and telecommunication relay. With the increasing demand for multiple UAV systems to complete complex tasks cooperatively, it is important to realize formation flight in a robust manner for these multi-agent systems. Multiple studies have been conducted to deal with the formation control problems in a team of unmanned vehicles.

In practical applications, it is difficult for a single UAV to perform complex tasks, due to the limited endurance, load capacity, and coverage radius. Distributed formation cooperative control can effectively organize multiple UAVs to accomplish various tasks, which effectively improve the operating radius and greatly enhance the application prospects of UAVs in various fields, such as border patrol, search and exploration, forest fire monitoring, and communication relay.

Rotary-wing UAV has the advantages of simple aerodynamic shape, vertical take-off and landing, low price, and so on, which is widely used in multiple fields. In recent years, rotary-wing UAV is also favored by military and civil fields. Because of its strong mobility, small size, and simple structure, rotary-wing UAV is used in the military field for indoor anti-terrorism reconnaissance and military strike and in the civil field for cluster performance, aerial photography, and news broadcast. In recent years, with the development of multi-agent research, rotary-wing is also used as a typical experimental platform for collaborative control algorithm research.

DOI: 10.1201/9781003242147-1

However, there are multiple challenges to the robust formation controller design of a group of UAVs. The dynamics of UAVs are highly nonlinear and coupled. Multiple uncertainties such as unmodeled uncertainties, parametric perturbations, and external disturbances can affect the tracking performance of the flight control system. Furthermore, in practical applications, such as source seeking and target enclosing, the UAV formation is required to track the desired trajectory to perform the tasks while keeping the desired time-varying formation. Moreover, individual UAV cooperates with others through communication for information sharing. During the information exchange, time delays are inevitable, which may degrade the control performance of the formation and even destabilize the entire system. In addition, each UAV may be subject to certain actuator failures. In the framework of formation control, the actuator failure in an individual UAV can spread over neighboring UAVs through the interaction topology and affect the performance of the whole system. All of the above issues make it difficult to achieve desired formation control performance for a group of UAVs. Therefore, the formation control for multiple UAVs needs to be solved first.

1.2 Literature Review on Formation Control for UAVs

1.2.1 UAV Formation Experiment

At present, researchers have made lots of achievements in UAV formation research, but most of them are mainly theoretical and simulation verification. In the actual formation flight experiment, we should not only consider the appropriate formation control method but also address the problems of information exchange, coordination, positioning and navigation, data calculation, and analysis among UAVs. This section mainly introduces the main research results of UAV formation flight experiments.

A group of YF-22 UAVs were designed and manufactured by the University of West Virginia for the experimental flight research to verify the formation control algorithm [1]. The formation control method was the classical leader-following method. In the formation flight experiment, the ground operator controlled the leader UAV through radio equipment, while the follower kept the formation and followed the position and direction of the leader (Figure 1.1).

Three quadrotor UAVs were used in the experiment of the University of Macau to perform the formation trajectory tracking flight experiment based on the leader-following strategy [2]. The formation experiment was based on the rapid prototyping and testing architecture in the MATLAB/Simulink environment, which integrated navigation data from a vicon motion capture system with 12 high-speed cameras, control algorithms, and communications with three radio-controlled quadrotors. During the experiment, the

FIGURE 1.1
Two YF-22 UAVs from the University of West Virginia perform leader-following formation [1].

leader quadrotor tracked the "8"-shaped trajectory and formed a triangle formation with the other two quadrotors to complete the formation mission and trajectory tracking.

Scholars from the School of Mechanical Engineering, Tokyo University, Japan, proposed a model prediction formation control method. Three small quadrotor UAVs were used for anti-collision formation control experiments [3]. In order to achieve formation control, the consensus control method was introduced into the model predictive control to maintain the formation of three quadrotors in the experiment. In the experiment, three-dimensional (3D) static cameras were used to obtain the position and attitude information of quadrotors in formation flight. All 3D static cameras were connected to one PC, which broadcasted the measurement data. Multiple PCs equaled to the number of UAVs were connected to the measurement PC through wireless network. Each controller obtained the current state information from the measurement PC and the predicted state information from other controllers to complete the distributed formation control.

The UAV research team of the Department of Electrical and Computer Science, National University of Singapore, has completed the formation experimental research on two unmanned helicopters named "Helion" and "Shelion" [4]. The formation adopted the classic leader-following formation strategy, "Helion" as the leader; "Sherion" received the leader's information to generate its own formation instructions. The communication between the two UAVs was completed through WiFi. The robust tracking control and leader-following formation control were combined to form the control method of formation system. The formation flight mission of two UAVs forming formation and tracking trajectory was completed (Figure 1.2).

FIGURE 1.2
Formation experiment of two unmanned helicopters of the National University of Singapore [4].

Kumar's team of the University of Pennsylvania in the United States completed the indoor formation flight experiment with multiple micro quadrotor UAVs [5]. In the formation experiment, a vicon motion capture system was used to complete the indoor positioning of quadrotor UAVs. The ground station obtained the positioning information, ran the formation controller, and sent the formation command to each quadrotor through ZigBee module to complete the formation flight control. The team accomplished the centralized cluster flight experiment, that is, the centralized processing of aircraft information was completed in the ground station, and the formation control command was sent to each aircraft from the ground station to complete the centralized formation flight. The team has successively realized the formation flight of 20 micro UAVs through vicon system navigation [6] and achieved the formation experiment of up to 33 micro UAVs' switching formation through ultra-wideband (UWB) indoor positioning technology [7].

The research group of LIS Laboratory of the Federal Institute of Technology in Lausanne, Switzerland, has realized the formation flight experiment of ten fixed-wing UAVs depended on swarm behavior [8]. The main purpose of the formation experiment was performed to study the influence of communication radius on flight stability. Each UAV completed autonomous flight navigation through a global positioning system (GPS) positioning, and each UAV communicated with each other through WiFi. Through the behavior-based formation strategy, the UAVs in the formation simulated the movement process of birds, fish, and other groups, and each UAV can only obtain the information of the surrounding individuals. This experiment verified that the formation strategy based on behavior was feasible, but there was no further research on other problems in formation flight (Figures 1.3–1.5).

FIGURE 1.3
University of Pennsylvania 20 Mini Quad rotor indoor formation flight [6].

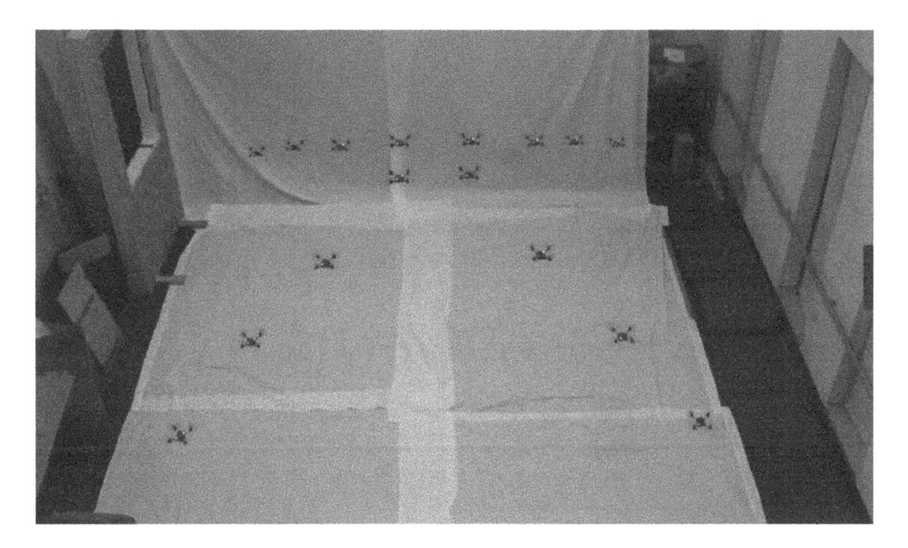

FIGURE 1.4
University of Pennsylvania 33 micro quadrotors switching formation flying [7].

South Korea's Gwangju Academy of Science and Technology has implemented a distributed formation flight experiment based on the distance using quadrotor UAVs [9]. In the formation experiment, the leading UAV flies at a constant reference speed, while the other following UAVs do not know the constant speed but use the adaptive method to estimate the reference speed. In formation, each UAV uses GPS positioning and its own inertial measurement unit data fusion to transform the global information into the local information of each aircraft through a rotation matrix. Each aircraft can

FIGURE 1.5
Flight experiment of ten fixed-wing UAV formation at Lausanne Federal Institute of Technology [8].

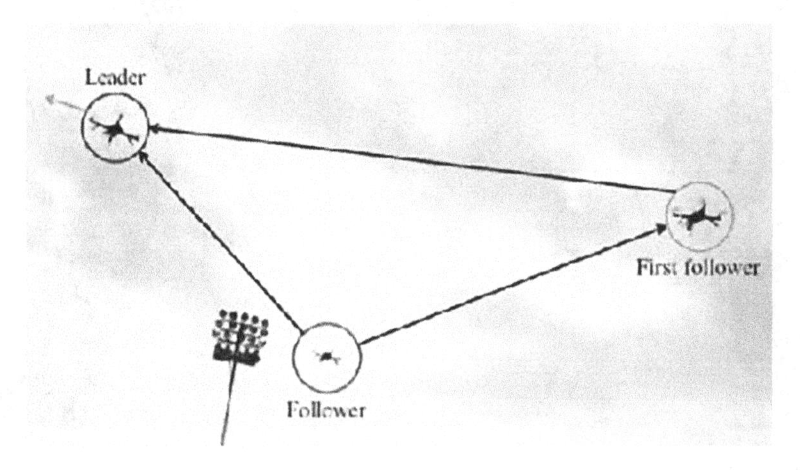

FIGURE 1.6
Distance-based distributed formation of three quadrotor UAVs of the Gwangju Institute of Science and Technology [9].

only obtain local information in formation flight, so as to realize distributed formation (Figure 1.6).

The team of School of Automation of Tsinghua University used the formation strategy based on consensus to realize the experiments of autonomous formation and time-varying formation of quadrotor UAVs [10]. In the experiment, a GPS system was used to locate the positioning of each UAV, ZigBee module was used to accomplish the wireless communication between UAVs, and a consensus formation controller was used to complete the formation flight of UAVs in different situations, such as fixed communication topology, switching topology, and time varying. The UAV team of School of Automation, Beijing University of Aeronautics and Astronautics, has completed the experimental study of multi-quadrotor UAV formation based on the switching topology method [11]. In the experiment, the leader received the reference trajectory command from the ground, and the followers formed a time-varying formation and followed the leader's state. Four quadrotor UAVs were used to complete the formation experiments under different conditions, such as time-varying formation, formation containment, and so on.

Using the formation control method based on the artificial potential field method, the UAV team of Harbin Institute of Technology carried out the formation flight experiment indoors and achieved the formation keeping, formation switching, and formation trajectory tracking [12]. In the experiment, the formation control command was sent to each UAV through the robot operating system, and the information exchange between adjacent UAVs was completed (Figures 1.7–1.9).

1.2.2 Research on UAV Formation Control Method

In recent decades, UAV formation control has been widely concerned by scholars. However, the UAV control method design is subjected to the following problems: first, in order to better match the characteristics of the UAV in the actual flight, the UAV dynamic equation should adopt the underactuated, nonlinear, and coupled model. Second, there are unmodeled dynamics, parameter perturbations, and external disturbances in the practical application of UAV. Finally, there are communication delays between UAVs and

FIGURE 1.7
The autonomous formation experiment of UAV in automation college of Tsinghua University [10].

FIGURE 1.8
Formation flight experiment of School of Automation, Beijing University of Aeronautics and Astronautics [11].

FIGURE 1.9
Formation experiment of Harbin Institute of Technology based on the artificial potential field method [12].

ground stations during formation flight. All these factors bring challenges to UAV formation control.

In [13–31], the cooperative formation control for a group of rotorcraft UAVs was investigated. In [13–16], formation control problems for a group of UAVs based on the leader-follower method were studied. In [13], a hybrid supervisory control method was proposed, which formed through a supervisory controller, maintained the achieved formation, and realized the formation flight of 3-degrees-of-freedom (3-DOF) UAVs. However, in the process of controller design, only the simplified translation dynamics were considered in the dynamic model, and the rotation dynamics were not considered.

A linear control law based on leader-follower formation was proposed in [14], which considered the translation and rotation dynamics. However, the nonlinear dynamics were simplified to linear systems by feedback linearization. A finite time formation control method was proposed in [15], but the complex rotational dynamics of each quadrotor in the formation was simplified. In [16], the suboptimal controller was proposed to solve the problem of parameter uncertainty and external disturbance suppression in the formation process. The UAV model of formation adopted a nonlinear model, but the stability of the whole closed loop under the drive formation system was not further proved, and the control method proposed cannot suppress the interference in the whole frequency domain. In [17], the behavior-based formation method was studied for the coordinated control of UAVs with priority tasks in obstacle areas. However, in the design of the formation controller, only translation dynamics was considered for each UAV, and the stability of the whole closed-loop system was not further analyzed. The control scheme based on potential field theory was discussed in [18, 19]. Among them, the UAV model was simplified to the second-order nonlinear model by feedback linearization. Furthermore, the authors proved the stability and regarded the formation UAV system as a large-scale interconnected linear system to calculate the value of the formation gain. In [20], the authors proposed formation control strategies based on navigation follow-up, behavior control, and virtual structure, which can be unified in the general framework based on the consensus control method. In [21–26], the formation coordination problem of quadrotors was investigated by using a formation strategy based on the consensus method. In [21], a complete nonlinear model was adopted to design the controller, but the stability analysis of the underactuated system was not complete. In [22, 23], a robust distributed formation control method for a group of 3-DOF helicopters was studied. However, in the research of trajectory tracking, the dynamics of the underactuated system with uncertainties was not considered. In [24], a distributed formation method of quadrotors based on nonsmooth second-order consensus control was proposed, but the stability of the nonlinear closed-loop system of the underactuated UAV was not proved. In [25], an adaptive formation control method based on the consensus strategy was proposed to solve the parameter perturbation problem in the whole closed-loop control system.

In addition, some methods to solve the problem of time delay in formation control are also mentioned in many literatures. In [26], a formation control method based on model predictive control was proposed, and the processing method of communication delay in formation was discussed. This control strategy can find a feasible approximate optimal control sequence with short and constant delay in the formation process. In [27], aiming at the communication delay problem of a class of underactuated systems, a formation method based on singularity free extraction algorithm was proposed, and the control schemes with constant communication delay and time-varying communication delay were given, respectively. In addition, the stability of the closed-loop system was proved by the Lyapunov method. In [28], a neighborhood feedback control method was proposed to reduce the communication delay between different individuals in the formation. In addition, it was proved that the formation consensus was asymptotically convergence, and an estimate of the convergence rate was given.

When performing more complex formation tasks, such as formation encirclement and target source search, compared with the time-invariant formation, the time-varying formation is more flexible and conducive to the implementation of complex tasks. In time-varying formation, because the communication distance between UAVs is limited, the method of switching communication topology is more conducive to the time-varying formation task. It was proposed in [29] that the design of the formation control method with time-varying formation and switching communication topology was more challenging than that with the fixed formation and fixed communication topology. In [30], a hierarchical controller based on backstepping control was proposed to realize the switching communication topology formation flying of a group of UAVs. However, the formation is partially time varying, that is, the velocity components of all UAVs must be the same in the formation process. However, in practical applications, many situations require different velocity components of UAV, such as rotating formation. In [31], the problem of time-varying formation based on consensus was studied. The necessary and sufficient conditions for the UAV group system to realize time-varying formation were given, and the explicit expression of time-varying formation center function was designed.

In the complex flight environment, several faults inevitably occur in UAVs, such as aging of electronic components, sensor failure, and actuator damage. These faults may reduce the stability of the control system. At present, several scholars have carried out works on fault-tolerant control technology. In [32], a fault-tolerant cooperative control strategy was proposed for a team of UAVs and unmanned ground vehicles in the presence of actuator faults. In [33], a fault-tolerant method for stabilization and navigation of 3D heterogeneous formations was proposed based on the model predictive control method. In [34], the problem of safe control for the trailing UAVs against actuator faults and input saturation was investigated. The external disturbances and internal actuator faults are estimated by using disturbance observers, and a backstepping control law was developed. In [35], a cooperative control problem for a multi-quadrotor system with quadrotor actuator faults was investigated. The extra cooperative controllers based on a potential-like function were used to make the faulty ones leave the formation without colliding with others.

Although scholars have done a lot of research on UAV formation control, UAVs in practical formation flight are subjected to nonlinear and coupling, parameter perturbation, communication delay, actuators faults, and external disturbances. It is significant to improve the robustness of the formation flight system by considering these factors simultaneously. Therefore, robust formation control methods and formation experiments for UAVs are worthy of further research.

1.3 Formation Platform

1.3.1 Introduction of Quadrotor Formation Hardware System

The diagram of the quadrotor used in this book is shown in Figure 1.10. The diagonal distance of the quadrotor is 92 mm, and the mass is 37 g.

During the flight, the four rotors of the quadrotor are distributed in an "X" shape. The two rotors at the lower left corner (No. 1) and the upper right corner (No. 3) rotate counterclockwise, and the two rotors at the upper left corner (No. 4) and the lower right corner (No. 2) rotate clockwise. When the sum of the lift generated by the rotation of the four rotors is equal to the gravity of the quadrotor, the quadrotor remains hovering. At the same time, the speed of four rotors increases or decreases to realize the vertical lifting movement. Rotors 1 and 2 accelerate (decelerate) and rotors 3 and 4 decelerate (accelerate) to realize the pitching motion of the quadrotor. Rotors 1 and 3 accelerate (decelerate) and rotors 2 and 4 decelerate (accelerate) to realize the yaw motion of the quadrotor. Rotors 1 and 4 accelerate (decelerate) and rotors 2 and 3 decelerate (accelerate) to realize the rolling motion of the quadrotor.

The quadrotor generates control input through four motors, but it has three translational degrees of freedom (laternal, longitudinal, and height) and three rotational degrees of freedom (pitch, yaw, and roll). Therefore, the quadrotor is an underactuated system with four inputs and six outputs.

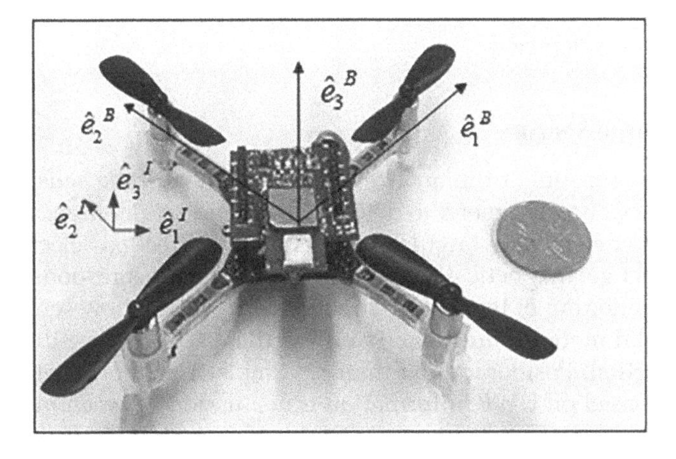

FIGURE 1.10
The diagram of a quadrotor.

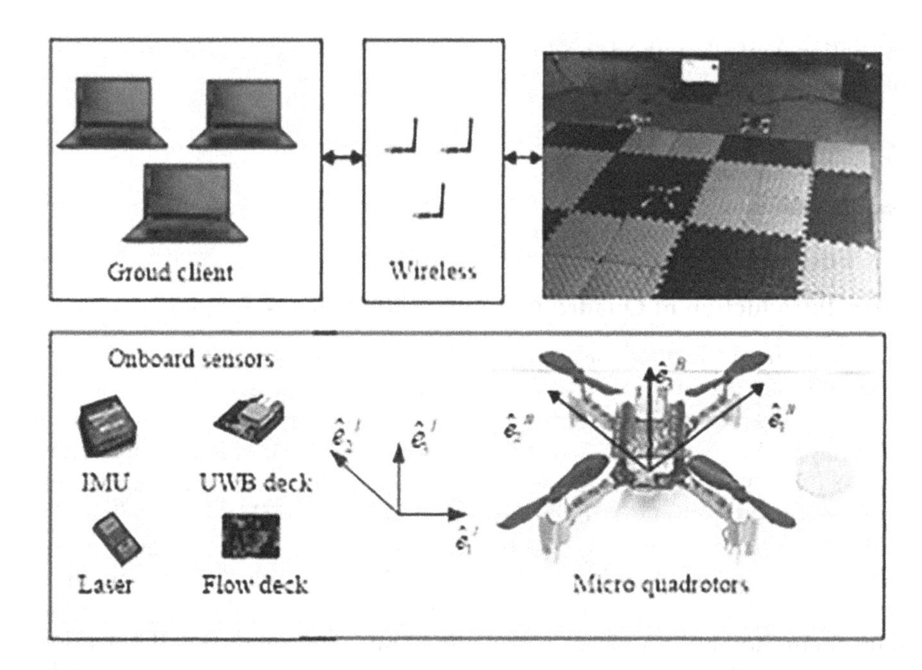

FIGURE 1.11
Hardware diagram of quadrotor formation.

In the formation of quadrotors, each quadrotor needs to obtain its own position, attitude, independent control, and communicate with neighbor quadrotors and ground station. In addition, the ground software needs to record the flight data of each UAV in the formation flight in real time. As shown in Figure 1.11, the hardware composition of the formation system of the quadrotor is depicted.

1.3.2 Airborne Sensors

The airborne sensors of the quadrotor are mainly attitude sensors, including a three-axis accelerometer to detect gravity and acceleration, a three-axis gyroscope to detect body angular velocity, and a three-axis electronic compass to detect geomagnetic direction. The laser sensor is responsible for the accurate positioning of the height direction. The optical flow sensor obtains the horizontal motion information and calculates the relative displacement in X–Y direction. Positioning tag mainly completes indoor positioning and navigation based on UWB, information transmission between tag and positioning base station, and indoor positioning of a quadrotor. The indoor positioning system based on UWB technology will be introduced in more detail in the next section.

1.3.3 Indoor Positioning System Based on UWB Technology

This monograph mainly completes the formation flight experiment of the quadrotor indoor; indoor positioning is the premise of the formation experiment of quadrotors. At present, there are many kinds of indoor positioning systems; this book mainly uses the indoor positioning system based on UWB technology to complete the indoor positioning of quadrotor formation.

The indoor positioning system based on UWB technology mainly includes positioning base station and positioning label. The positioning base station is built in the field where a formation flight experiment is needed, and the quadrotor carries the positioning tag, which is used to complete the communication between the positioning tag and the positioning base station. Figure 1.12 is the schematic diagram of the construction of the experimental positioning system. Six positioning base stations are arranged in the indoor environment, and the time difference of arrival (TDOA) algorithm is adopted [27]. The distance between the positioning tag carried by the quadrotor and different base stations is different, and the time for the signal sent by the same tag to reach different base stations is different, resulting in the "time difference of arrival" of the arrival signal at different times. In this book, the TDOA of the tag signal received by different base stations is used to calculate the location of the positioning tag.

In the part of this experiment, due to the influence of UWB positioning technology, such as positioning tag and random interference of base station communication, the positioning coordinates of quadrotor calculated

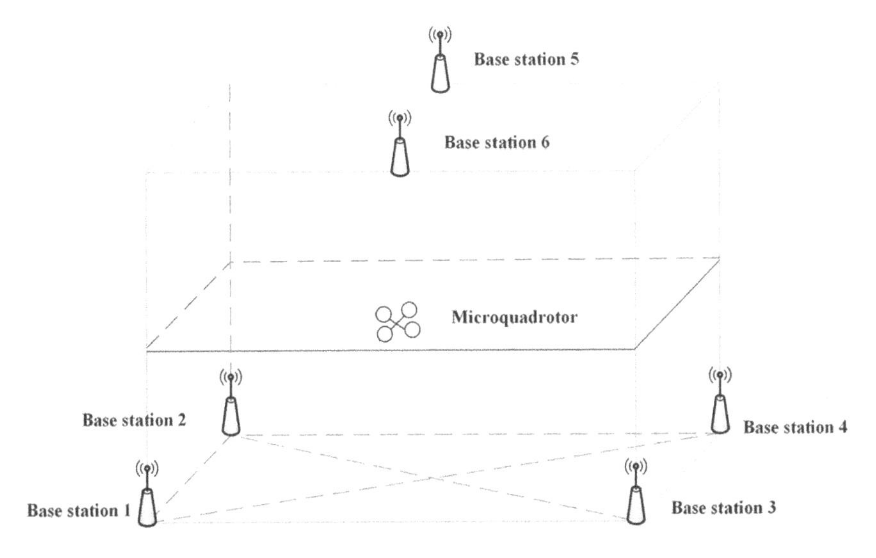

FIGURE 1.12
Schematic diagram of an indoor positioning system.

FIGURE 1.13
Schematic diagram of the indoor positioning system in the experiment.

have errors, which affect the precise positioning and navigation of quadrotor. In order to improve the positioning accuracy in the formation process, the positioning and navigation of the quadrotor formation process are completed by the indoor positioning system based on UWB technology and optical flow and laser module, and three parts of data fusion are completed by a Kalman filter and a complementary filter, so that the positioning accuracy of the quadrotor formation process is more accurate, reaching centimeter-level positioning error. The positioning system built during the experiment is shown in Figure 1.13. White circles indicate the location of each positioning base station.

1.3.4 Communication Module

Due to the distributed formation task, each UAV in the formation cannot obtain the global information but obtain its own formation instructions and the formation information of its neighbors. The UAV communicates with the ground station and its neighbors through the 2.4 GHz wireless communication module. During the flight, the UAV can send real-time position information to the ground station through the wireless communication module, and the communication network between the UAVs can also be completed through this communication module.

1.4 Preview of Chapters

In this monograph, we address the formation control problems, including time-varying formation, communication delays, fault-tolerant formation, for multiple UAV systems with highly nonlinear and coupled parameter uncertainties, and external disturbances. Differentiating from existing works, we present a robust optimal formation approach to design distributed cooperative control laws for a group of UAVs, based on the linear quadratic regulator control method and the robust compensation theory. The proposed control method is composed of two parts: the nominal part to achieve desired tracking performance and the robust compensation part to restrain the influence of highly nonlinear and strongly coupled parameter uncertainties and external disturbances on the global closed-loop control system. Furthermore, this book gives the proof of robust properties. The influence of communication delays and actuator fault tolerance can be restrained by the proposed robust formation control protocol, and the formation tracking errors can converge into a neighborhood of the origin bounded by a given constant in a finite time. Moreover, this book provides details about the practical application of the proposed method to design formation control systems for multiple quadrotors and tail sitters.

This monograph consists of nine chapters. A brief introduction of these chapters is given as follows.

Chapter 1 introduces the background of formation control for multiple UAVs and the research of UAV formation control experiments and control methods, and the chapter arrangement of this book.

Chapter 2 investigates the robust formation control problem for a group of quadrotors subject to underactuation, high nonlinearities, and couplings. A distributed robust controller is developed. Theoretical analysis and simulation studies of the formation of multiple uncertain quadrotors are also presented to validate the effectiveness of the proposed formation control scheme.

Chapter 3, a continuation of Chapter 2, studies the robust formation control problem for a team of quadrotors with communication delays. A distributed formation controller is designed for the quadrotor team. It's further proven that the tracking errors can converge into a neighborhood of the origin bounded by a given constant in a finite time. Experimental results on multiple quadrotors are provided to verify the effectiveness of the proposed control method.

Chapter 4 focuses on the formation control for a team of quadrotors subject to switching topologies. A robust control approach is proposed that consists of a position and an attitude controller. Theoretical foundations, detailed simulation studies, and experimental results validate the effectiveness of the proposed methodology.

Chapter 5 considers the time-varying formation control problem or a group of tail sitters in flight mode transitions between forward and vertical flight. A robust distributed formation control method is proposed to achieve the aggressive time-varying formation subject to nonlinear dynamics and uncertainties. The robustness issue of the robust protocols is also discussed. Simulation results for a team of tail sitters to achieve the time-varying formation in flight mode transitions are provided to show the effectiveness of the proposed control strategy.

Chapter 6 studies the fault-tolerant time-varying formation control problem for a group of tail sitters with multiple actuator faults and uncertainties. A robust distributed fault-tolerant formation control strategy is developed. The information of the actuator faults does not need to be identified online and the tracking errors of the global closed-loop control system can converge into a given neighborhood of the origin in a finite time. Simulation results are also presented to show the effectiveness of the proposed control strategy.

2

Robust Formation Control for Multiple Quadrotors with Nonlinearities and Disturbances

Formation control for a group of unmanned aerial vehicles (UAVs) is challenging due to the complex systems dynamics in the cooperative feedback loops. This chapter investigates the problem of the robust formation control for a group of quadrotors with underactuation, high nonlinearities and couplings, and disturbances. A distributed robust controller is developed, which consists of a position controller to govern the translational motion for the desired formation and an attitude controller to control the rotational motion of each quadrotor. Both rigorous theoretical analysis and simulation results are presented to validate the effectiveness of this method.

2.1 Introduction

Formation control of multiple autonomous UAVs has received much recent attention because of potentially broad applications in engineering and science, such as robotics, aerospace, and wireless communication, as illustrated in [36–39]. Multiple UAVs can be used to carry out tasks in dangerous situations, for example, reconnaissancing over hostile territories or tracking battle damage of enemy targets (see, e.g., [36–39]). UAVs have many potential military and civil applications and are also of great scientific significance in academic research (see [40, 41] and the references therein). With the increasing demand of multi-UAV systems to cooperate and complete complex tasks, it is important to realize flight formation in a robust manner for these multiagent systems.

Motivated by the wide range of possible military and civilian applications, formation control of UAVs has attracted a considerable amount of research interest in the past decade. According to different criteria, formation control approaches in the existing literature have different classifications. In the last two decades, several classical methodologies have been developed to address the formation control of multi-agent systems, including the behavioral strategy, the virtual structure method, and the leader-follower approach

DOI: 10.1201/9781003242147-2

(see, e.g., [2, 14–16, 19, 42–45]). Each of the above three approaches has disadvantages [46]. For example, it is hard for the behavioral strategy to explicitly define the group behavior or mathematically analyze the stability properties of the global formation system. The virtual structure-based strategy requires the formation to act as a virtual structure and limits the potential application scales of this approach. Furthermore, there is no explicit feedback from the followers to the leaders, and the leader-follower method loses robustness upon the failure of the leader.

Recently, consensus problems of linear time-invariant swarm systems or multi-agent systems have been studied extensively. The leader-follower, behavioral, and virtual structure-based formation control methods can all be unified in this general framework [19]. Many results that utilize the consensus protocols for the UAV formation control have been derived in, e.g., [11, 21–24]. Quadrotors serve as efficient aerial platforms and possess many features such as the capabilities to hover and take off and land vertically and have simple mechanisms without swashplates or linkages, as depicted in [48, 49].

In recent years, many studies on formation control of UAVs have been carried out in both aeronautics and robotics communities. In [11, 26, 39, 43, 44], each vehicle system was simplified as a first-order or second-order linear system and the rotational dynamics was ignored in the stability analysis of the multi-agent systems. In [19, 22, 23, 38], the formation control problems were addressed for multiple nonlinear systems. In [2, 14, 15, 45], the cooperative tracking problems were discussed for a group of quadrotors with translational and rotational motions, but the complex nonlinear vehicle dynamics was simplified or linearized. In practical applications, the quadrotor is a multiple-input multiple-output underactuated system involving highly nonlinear and strongly coupled dynamics. The nonlinear models of complex dynamics and 6-degrees-of-freedom (6-DOF) were also considered in the flight formation control of multiple quadrotors. In [16], a robust formation control method based on leader-follower structure was proposed for quadrotors with nonlinear dynamics. In [21, 44], the consensus-based formation control strategies were constructed for multiple nonlinear quadrotor systems. However, the disturbance rejection problems were ignored in many previously published studies on the design of flight formation controllers. In formation flight, environmental uncertainties such as atmospheric disturbances act as additional forces and moments on the quadrotor dynamics, which can seriously affect the stability of the quadrotor formation system.

In [22, 23], a robust distributed formation controller was designed for a class of 3DOF helicopters to achieve desired formation flying under external disturbances. In [25], an adaptive consensus control method was developed for multiple nonlinear agent systems. However, the underacturated and uncertain 6DOF system dynamics were not further considered. In [16], a suboptimal control controller was designed for the classical leader-follower formation control problem of uncertain quadrotors; however, the stability for the overall underacturated multi-UAV systems was not fully analyzed and the control method cannot reject the disturbances in the whole frequency range as much as desired.

Motivated by the above factors, the formation control for multiple quadrotor swarm systems subject to disturbances and underactuated, highly nonlinear, and strongly coupled agent dynamics is studied in this chapter. A robust control approach is developed to address the formation control problem for a group of quadrotors with nonlinearities and uncertainties. The proposed overall controller consists of a position controller to govern the translational dynamics to achieve the desired formation trajectory and pattern and an attitude controller design to stabilize the inner rotational dynamics. Compared to previous studies on quadrotor formation control, the main contribution of this chapter is threefold, which is summarized as follows.

First, each quadrotor considered here is underactuated. However, the formation control problems for such underactuated systems were not fully studied in [11, 22, 23, 26, 38, 39, 43–45]. Second, the external disturbances are considered in both the translational and rotational dynamics of the underactuated model. In [36, 38, 41] and [2, 11, 15, 21, 27, 45, 48], the disturbance restraining problem was ignored, while the stability of the whole uncertain and nonlinear underactuated multi-UAV system was not fully analyzed in [15, 16, 22–25]. Third, highly nonlinear and strongly coupled dynamics are considered, and the formation control problem under aggressive maneuvers is addressed for a group of quadrotors. In particular, the translational dynamics is bilinear, and the backstepping approach newly developed in [50] is used to obtain an inverse kinematrics solution of the quadrotor model. However, the formation control problem is not limited to the hovering or low-speed scenarios that are constrained by the simplified or linearized vehicle model, as adopted in [2, 15, 45]. In addition, the graph theory is introduced to address the consensus-based formation control problem and the resulting global flight control laws are distributed.

2.2 Preliminaries and Problem Formulation

In this section, basic concepts of the graph theory are introduced and the problem description is presented.

2.2.1 Quadrotor Model

Consider the dynamical model of quadrotor i as a rigid body. Let $p=\begin{bmatrix}p_x & p_y & p_z\end{bmatrix}^T \in \mathbb{R}^{3\times 1}$ represent the position vectors in the earth-fixed inertial coordinate and $\Theta=\begin{bmatrix}\phi & \theta & \psi\end{bmatrix}^T \in \mathbb{R}^{3\times 1}$ indicate the three Euler angles, i.e., the roll, pitch, and yaw angles, respectively. The roll and pitch angles of the ith quadrotor are assumed to satisfy: $|\phi_i| < \pi/2$, $|\theta_i| < \pi/2$, and $|\psi_i| < \pi/2$, respectively, to avoid the singularity problem in the Euler angle expressions. From [51], the full dynamics of a quadrotor can be modeled as

$$m\ddot{p} = Rf,$$

$$J\ddot{\Theta} = C(\Theta,\dot{\Theta})\dot{\Theta} + \tau, \tag{2.1}$$

where m denotes the quadrotor mass and $J = diag\{J_\phi, J_\theta, J_\psi\} \in \mathbb{R}^{3\times3}$ denotes the inertia matrix. Denote $R \in SO(3)$ as the rotation matrix, which can be described as

$$R = \begin{bmatrix} \cos\theta\cos\psi & \cos\psi\sin\phi\sin\theta - \cos\phi\sin\psi & \sin\phi\sin\psi + \cos\phi\cos\psi\sin\theta \\ \cos\theta\sin\psi & \cos\phi\cos\psi + \sin\phi\sin\theta\sin\psi & \cos\phi\sin\theta\sin\psi - \cos\psi\sin\phi \\ -\sin\theta & \cos\theta\sin\phi & \cos\phi\cos\theta \end{bmatrix}.$$

As depicted in [51], the Coriolis term $C(\Theta,\dot{\Theta}) = [c_{ij}] \in \mathbb{R}^{3\times3}$ can be described as follows:

$$c_{11} = 0,$$

$$c_{12} = (J_\theta - J_\psi)(\dot{\theta}\cos\phi\sin\phi + \dot{\psi}\sin^2\phi\cos\theta) + (J_\psi - J_\theta)\dot{\psi}\cos^2\phi\cos\theta - \dot{\psi}\cos\theta,$$

$$c_{13} = (J_\psi - J_\theta)\dot{\psi}\cos\phi\sin\phi\cos^2\theta,$$

$$c_{21} = (J_\psi - J_\theta)(\dot{\theta}\cos\phi\sin\phi + \dot{\psi}\sin^2\phi\cos\theta) + (J_\theta - J_\psi)\dot{\psi}\cos^2\phi\cos\theta + J_\phi\dot{\psi}\cos\theta,$$

$$c_{22} = (J_\psi - J_\theta)\dot{\phi}\cos\phi\sin\phi,$$

$$c_{23} = -J_\phi\dot{\psi}\sin\theta\cos\theta + J_\psi\dot{\psi}\cos^2\phi\cos\theta\sin\theta + J_\theta\dot{\psi}\sin^2\phi\cos\theta\sin\theta,$$

$$c_{31} = (J_\theta - J_\psi)\dot{\psi}\cos^2\theta\sin\phi\cos\phi - J_\phi\dot{\theta}\cos\theta,$$

$$c_{32} = (J_\psi - J_\theta)(\dot{\theta}\cos\phi\sin\phi\sin\theta + \dot{\phi}\sin^2\phi\cos\theta) - J_\theta\dot{\psi}\sin^2\phi\sin\theta\cos\theta$$
$$+ (J_\theta - J_\psi)\dot{\phi}\cos^2\phi\cos\theta + J_\phi\dot{\psi}\sin\theta\cos\theta - J_\psi\dot{\psi}\cos^2\phi\cos\theta\sin\theta,$$

$$c_{33} = (J_\theta - J_\psi)\dot{\phi}\cos\phi\sin\phi\cos^2\theta + J_\phi\dot{\theta}\sin\theta\cos\theta - J_\theta\dot{\theta}\sin^2\phi\cos\theta\sin\theta$$
$$- J_\psi\dot{\theta}\cos^2\phi\sin\theta\cos\theta.$$

The external force f of the quadrotor can be calculated by

$$f = \begin{bmatrix} 0 & 0 & f_T \end{bmatrix}^T - R^T \begin{bmatrix} 0 & 0 & mg \end{bmatrix}^T, \tag{2.2}$$

where g represents the gravity constant. The total lift f_T and the torque $\tau = \begin{bmatrix} \tau_\phi & \tau_\theta & \tau_\psi \end{bmatrix}^T$ can be expressed as

$$\begin{aligned} f_T &= k_f \sum_{j=1}^{4} \omega_j^2, \\ \tau_\phi &= l_c k_f (\omega_2^2 - \omega_4^2), \\ \tau_\theta &= l_c k_f (\omega_1^2 - \omega_3^2), \\ \tau_\psi &= k_\tau \sum_{j=1}^{4} (-1)^{j+1} \omega_j^2, \end{aligned} \tag{2.3}$$

where k_f, l_c, k_τ are positive parameters and ω_j $(j = 1,2,3,4)$ are the rotational speeds of Rotor j. The control input commands to the four rotors can be written as follows:

$$\begin{aligned} u_z &= \sum_{j=1}^{4} \omega_j^2, \\ u_\phi &= \omega_2^2 - \omega_4^2, \\ u_\theta &= \omega_1^2 - \omega_3^2, \\ u_\psi &= \sum_{j=1}^{4} (-1)^{j+1} \omega_j^2. \end{aligned} \tag{2.4}$$

Remark 2.1

It should be pointed out that quadrotor i is an underactuated system, which possesses 6-DOF (the height, the lateral and longitudinal positions, and three attitude angles) and four inputs (u_{zi}, $u_{\Theta 1,i}$, $u_{\Theta 2,i}$, $u_{\Theta 3,i}$). In this chapter, the position and heading control mode is chosen for each quadrotor, including the longitudinal position p_x, the lateral position p_y, the height p_z, and the yaw angle ψ as outputs and u_θ, u_ϕ, u_z, and u_ψ as control inputs.

Remark 2.2

It can be observed that the vehicle system is highly nonlinear and strongly coupled, especially considering the expressions of the Coriolis term $C(\Theta, \dot{\Theta})$.

2.2.2 Preliminaries on Graph Theory

Consider a group of N quadrotors labeled from 1 to N. Let $\Phi = \{1,2,\ldots,N\}$. Let directed graph $G = (V,E,W)$ represents the information exchange among the N quadrotors. $V = \{v_1, v_2, \ldots, v_N\}$ denotes the set of nodes, where v_i denotes the ith quadrotor. $E \subseteq \{(v_i,v_j): v_i,v_j \in V, i \neq j\}$ denotes the set of edges, and $\varepsilon_{ij} = (v_i,v_j) \in E$ indicates that the ith quadrotor can receive information from its neighbor, the jth quadrotor. Let $N_i = \{j | (v_i,v_j) \in E\}$ denote the set of neighbors of a node v_i. $W = [w_{ij}] \in \mathbb{R}^{N\times N}$ is the weighted adjacency matrix associated with G. For any $i, j \in \Phi$, $w_{ij} > 0$ if and only if $\varepsilon_{ij} \in E$, and $w_{ij} = 0$ otherwise. A simple graph is considered here and thereby $w_{ii} = 0$. For the node v_i, the weighted in-degree d_i is defined as the ith row sum of W as $d_i = \sum_{j=1}^{N} w_{ij}$. Define the in-degree matrix as $D = diag\{d_i\}$ and the weighted graph Laplacian matrix as $L = D - W$. A directed path from v_{i_1} to v_{i_r} is termed as a sequence of ordered edges in the form $\varepsilon_{i_k i_{k+1}} (k = 1,2,\ldots,r-1)$. The graph G is said to contain a spanning tree if there exists a root node having directed paths to all other nodes, and the node is called the root.

2.2.3 Problem Formulation

The main goal of this chapter is to design a distributed robust controller for the group of quadrotors to achieve the desired formation trajectory while keeping the desired formation pattern. The prescribed reference trajectory of the formation center is denoted as $p_{r0} \in \mathbb{R}^{3\times1}$, which can also be viewed as the trajectory of a virtual group leader. The reference p_{r0} is assumed to be differentiable and its second derivative $\ddot{p}_{r0} = 0$. Let $\delta_{ij} = \delta_i - \delta_j$, where δ_i or δ_j can represent the desired position deviations between the formation center p_{r0} and the ith quadrotor or the jth quadrotor, respectively. In this chapter, the time-invariant formation pattern is considered for the group of quadrotors, and thereby δ_{ij} is a constant. Define the desired position deviation between the ith quadrotor and the jth quadrotor as $\delta_{ij} = \begin{bmatrix} \delta_{x,ij} & \delta_{y,ij} & \delta_{z,ij} \end{bmatrix}^T \in \mathbb{R}^{3\times1}$ $(i, j \in \Phi)$. δ_{ij} determines the formation pattern of the quadrotor group.

Then, for the ith quadrotor, the translational and rotational model can be rewritten as follows:

$$\begin{aligned} \ddot{p}_i &= u_{zi} B_{pi} R_i c_{3,3} - g c_{3,3}, \\ \ddot{\Theta}_i &= J_i^{-1} C(\Theta_i, \dot{\Theta}_i) \dot{\Theta}_i + B_{\Theta i} u_{\Theta i}, \end{aligned} \tag{2.5}$$

respectively, where $u_{\Theta i} = \begin{bmatrix} u_{\phi i} & u_{\theta i} & u_{\psi i} \end{bmatrix}^T$, $B_{\Theta i} = J_i^{-1} diag\{l_{ci}k_{fi}, l_{ci}k_{fi}, k_{\tau i}\}$, $B_{pi} = m_i^{-1}k_{fi}I_3$, and $c_{3,3} = \begin{bmatrix} 0 & 0 & 1 \end{bmatrix}^T$. For the ith quadrotor, one can obtain the following model by introducing disturbances:

$$\begin{aligned} \ddot{p}_i &= B_{pi}F_i - gc_{3,3} + B_{pi}\tilde{F}_i + d_{pi}, \\ \ddot{\Theta}_i &= J_i^{-1}C_i(\Theta_i, \dot{\Theta}_i)\dot{\Theta}_i + B_{\Theta i}u_{\Theta i} + d_{\Theta i}, \end{aligned} \tag{2.6}$$

where $\tilde{F}_i = u_{zi}R_ic_{3,3} - F_i$ and the atmospheric disturbances $d_{pi} \in \mathbb{R}^{3\times1}$ and $d_{\Theta i} \in \mathbb{R}^{3\times1}$ are additional forces and moments and are assumed to be bounded. The variable $F_i = \begin{bmatrix} F_{ji} \end{bmatrix} \in R^{3\times1}$ is the virtual position control input to be determined and satisfies

$$F_i = u_{zi}\begin{bmatrix} \sin\phi_i \sin\psi_i + \cos\phi_i \cos\psi_i \sin\theta_{ri} \\ \cos\phi_i \sin\theta_i \sin\psi_i - \cos\psi_i \sin\phi_{ri} \\ \cos\phi_i \cos\theta_i \end{bmatrix}, \tag{2.7}$$

where θ_{ri} and ϕ_{ri} are the references for the pitch and roll angles, respectively. The yaw angle is required to track the reference ψ_{ri}.

Remark 2.3

Because each quadrotor system is underactuated, the quadrotor dynamics is decomposed into the attitude dynamics and the position dynamics as given in the first and second equations in (2.6) for the backstepping-based controller design. Then, by considering F_i $(i \in \Phi)$ as the virtual position control input, the attitude and position controllers can be designed, respectively, based on the backstepping approach [50].

Define the uncertainties $\Delta_{pi}, \Delta_{\Theta i}$ as

$$\begin{aligned} \Delta_{pi} &= B_{pi}\tilde{F}_i + d_{pi}, \\ \Delta_{\Theta i} &= d_{\Theta i}, \ i \in \Phi. \end{aligned} \tag{2.8}$$

Here, $\Delta_{pi} = \begin{bmatrix} \Delta_{pj,i} \end{bmatrix} \in \mathbb{R}^{3\times1}$ and $\Delta_{\Theta i} = \begin{bmatrix} \Delta_{\Theta j,i} \end{bmatrix} \in \mathbb{R}^{3\times1}$ are the equivalent disturbances involving the force error $\tilde{F}_i$. The nonlinear model (2.6) can then be rewritten as

$$\begin{aligned} \ddot{p}_i &= B_{pi}F_i - gc_{3,3} + \Delta_{pi}, \\ \ddot{\Theta}_i &= J_i^{-1}C_i(\Theta_i, \dot{\Theta}_i)\dot{\Theta}_i + B_{\Theta i}u_{\Theta i} + \Delta_{\Theta i}, \ i \in \Phi. \end{aligned} \tag{2.9}$$

Remark 2.4

One can see that the translational dynamics is bilinear due to the first term on the right-hand side, i.e., F_i $(i \in \Phi)$ from the first equation in (2.6). In fact, the dynamics of each quadrotor cannot simply be separated into the translational dynamics and the rotational dynamics, due to the existence of the coupled term $\tilde{F}_i$ in the first equation of (2.6). Actually, the force error $\tilde{F}_i$ is included in the equivalent disturbance Δ_{pi}.

Remark 2.5

The model (2.9) represents the real dynamical model of each quadrotor. One can obtain the nominal model by removing the equivalent disturbances Δ_{pi} and $\Delta_{\Theta i}$. The complete model includes the nominal model and equivalent disturbances.

2.3 Formation Protocol Design and System Analysis

In this section, the formation controller is designed based on the robust compensation theory, followed by the analysis of the tracking performance and the robustness property of the whole cooperative system. The designed controller consists of a position controller to govern the translational motion for the desired formation and an attitude controller to control the rotational motion of each quadrotor.

2.3.1 Position Controller Design

The virtual position control input F_i consists of two parts:

$$F_i = F_i^N + F_i^R,\ i \in \Phi, \tag{2.10}$$

where $F_i^N \in \mathbb{R}^{3\times 1}$ denotes the nominal part and $F_i^R \in \mathbb{R}^{3\times 1}$ denotes the robust compensating part.

The nominal control input $F_i^N \in \mathbb{R}^{3\times 1}$ is designed to achieve the desired formation control for the nominal translational system, while the robust compensating input $F_i^R \in \mathbb{R}^{3\times 1}$ is designed to restrain the effects of Δ_{pi} on the whole closed-loop system. The nominal part F_i^N $(i \in \Phi)$ is constructed by ignoring Δ_{pi} as follows:

$$\begin{aligned} F_i^N = & -\alpha_p \sum_{j \in N_i} w_{ij} B_{pi}^{-1} \left(K_p \left(p_i - p_j - \delta_{ij} \right) + K_v \left(\dot{p}_i - \dot{p}_j \right) \right) \\ & - \alpha_p b_{li} B_{pi}^{-1} \left(K_p \left(p_i - \delta_i - p_{r0} \right) + K_v \left(\dot{p}_i - \dot{p}_{r0} \right) \right) + g B_{pi}^{-1} c_{3,3}, \end{aligned} \tag{2.11}$$

where α_p represents a positive coupling gain and $K_p, K_v \in \mathbb{R}^{3\times3}$ are nominal controller parameter matrices with positive elements. The constant b_{li} indicates the connection weight between the virtual group leader and ith quadrotor: $b_{li} > 0$ represents that the ith quadrotor can obtain the information from the virtual leader, and $b_{li} = 0$ otherwise.

Furthermore, F_i^R $(i \in \Phi)$ is introduced to restrain the effects of Δ_{pi} on the real system. The robust compensating input is designed as follows:

$$F_i^R = \Gamma_{pi}(s)F_i^{R*} = -B_{pi}^{-1}\Gamma_{pi}(s)\Delta_{pi}. \tag{2.12}$$

where $\Gamma_{pi}(s) = diag\{\Gamma_{p1,i}(s), \Gamma_{p2,i}(s), \Gamma_{p3,i}(s)\}$ and $\Gamma_{pj,i}(s) = \eta_{pj,i}^2/(s+\eta_{pj,i})^2 (j = 1,2,3)$ with $\eta_{pj,i}$ $(j = 1,2,3)$ are positive robust filter parameters.

Remark 2.6

One can see that the robust filter $\Gamma_{pj,i}(s)$ has the following property. A larger $\eta_{pj,i}$ leads to a wider frequency bandwidth of $\Gamma_{pj,i}(s)$, within which the filter gain is closer to 1. In this case, one can see that the robust compensating input is approximate to $F_i^{R*} = -B_{pi}^{-1}\Delta_{pi}$; the effects of Δ_{pi} on the real system can be counteracted completely.

Remark 2.7

It should be noted that the equivalent disturbance Δ_{pi} $(i \in \Phi)$ does not represent a real external signal but includes the system state variables, which poses a challenge in the analysis of the tracking performances and robustness properties for the overall system.

The robust compensating input (2.12) cannot be implemented in practical applications because Δ_{pi} cannot be measured directly. From (2.9), one has that

$$\Delta_{pi} = \ddot{p}_i - B_{pi}F_i + gc_{3,3}, \; i \in \Phi. \tag{2.13}$$

Let $\eta_{pi} = diag\{\eta_{p1,i}, \eta_{p2,i}, \eta_{p3,i}\}$. Substituting (2.13) into (2.12), one can obtain that as

$$\begin{aligned}
\dot{\vartheta}_{1i}^p &= -\eta_{pi}\vartheta_{1i}^p - \eta_{pi}^2 p_i + B_{pi}F_i - gc_{3,3}, \\
\dot{\vartheta}_{2i}^p &= -\eta_{pi}\vartheta_{2i}^p + 2\eta_{pi}p_i + \vartheta_{1i}^p, \\
F_i^R &= B_{pi}^{-1}\eta_{pi}^2(\vartheta_{2i}^p - p_i), \; i \in \Phi.
\end{aligned} \tag{2.14}$$

Furthermore, by solving the equation in (2.7), one can obtain the vertical control input u_{zi} and the pitch and roll angle references θ_{ri} and ϕ_{ri} as follows:

$$u_{zi} = F_{3i}/\cos\phi_i/\cos\theta_i,$$
$$\theta_{ri} = \sin^{-1}\left((F_{1i}/u_{zi} - \sin\phi_i \sin\psi_i)/\cos\phi_i/\cos\psi_i)\right), \tag{2.15}$$
$$\phi_{ri} = \sin^{-1}\left((\cos\phi_i \sin\theta_i \sin\psi_i - F_{2i}/u_{zi})/\cos\psi_i\right).$$

2.3.2 Attitude Controller Design

The attitude controller is designed to achieve the desired attitude reference tracking. Let $\Theta_{ri} = \begin{bmatrix}\theta_{ri} & \phi_{ri} & \psi_{ri}\end{bmatrix}^T$ be the attitude reference and $e_{\Theta i} = \begin{bmatrix}e_{\phi i} & e_{\theta i} & e_{\psi i}\end{bmatrix}^T = \Theta_i - \Theta_{ri}$ $(i \in \Phi)$ be the attitude error. Similarly to the virtual position control input F_i, the attitude control input $u_{\Theta i}$ is constructed as follows:

$$u_{\Theta i} = u_{\Theta i}^N + u_{\Theta i}^R,\ i \in \Phi, \tag{2.16}$$

where $u_{\Theta i}^N$ denotes the nominal control part and $u_{\Theta i}^R$ denotes the robust compensating part. The nominal attitude control input $u_{\Theta i}^N$ can be designed as

$$u_{\Theta i}^N = B_{\Theta i}^{-1}\left(-K_\Theta e_{\Theta i} - K_\omega \dot{e}_{\Theta i} - J_i^{-1}C_i(\Theta_i, \dot{\Theta}_i)\dot{\Theta}_i + \ddot{\Theta}_{ri}\right). \tag{2.17}$$

The robust attitude compensating input $u_{\Theta i}^R$ is designed as

$$u_{\Theta i}^R = -B_{\Theta i}^{-1}\Gamma_{\Theta i}(s)\Delta_{\Theta i}, \tag{2.18}$$

where $K_\Theta, K_\omega \in \mathbb{R}^{3\times 3}$ are diagonal nominal controller parameter matrices with positive elements and $\Gamma_{\Theta i}(s) = diag\{\Gamma_{\Theta 1,i}(s), \Gamma_{\Theta 2,i}(s), \Gamma_{\Theta 3,i}(s)\}$, and $\Gamma_{\Theta j,i}(s) = \eta_{\Theta j,i}^2/(s+\eta_{\Theta j,i})^2$ are robust filters with positive parameters $\eta_{\Theta j,i}$ $(j = 1,2,3)$ to be determined. Let $\eta_{\Theta i} = diag\{\eta_{\Theta 1,i}, \eta_{\Theta 2,i}, \eta_{\Theta 3,i}\}$. Similarly, from (2.9) and (2.18), $u_{\Theta i}^R$ can be realized with filter states $\vartheta_{1i}^\Theta, \vartheta_{2i}^\Theta \in \mathbb{R}^{3\times 1}$ as follows:

$$\dot{\vartheta}_{1i}^\Theta = -\eta_{\Theta i}\vartheta_{1i}^\Theta - \eta_{\Theta i}^2\Theta_i + B_{\Theta i}u_{\Theta i} + J_i^{-1}C_i(\Theta_i, \dot{\Theta}_i)\dot{\Theta}_i,$$
$$\dot{\vartheta}_{2i}^\Theta = -\eta_{\Theta i}\vartheta_{2i}^\Theta + 2\eta_{\Theta i}\Theta_i + \vartheta_{1i}^\Theta, \tag{2.19}$$
$$u_{\Theta i}^R = B_{\Theta i}^{-1}\eta_{\Theta i}^2(\vartheta_{2i}^\Theta - \Theta_i),\ i \in \Phi.$$

Remark 2.8

One can see that the proposed robust formation controller is distributed and time invariant, which means that the controller of quadrotor i only depends on the position and velocity information from itself and its neighbors.

2.3.3 System Analysis

Define the position error as $e_{pi} = p_i - \delta_i - p_{r0} = \left[e_{pj,i} \right] \in \mathbb{R}^{3\times 1}$ $(i \in \Phi)$. Combining (2.9)–(2.11), one can obtain that

$$\begin{aligned} \ddot{p}_i = & -\alpha_p \sum_{j\in N_i} w_{ij}\left(K_p\left(e_{pi} - e_{pj}\right) + K_v\left(\dot{e}_{pi} - \dot{e}_{pj}\right)\right) \\ & - \alpha_p b_{li}\left(K_p e_{pi} + K_v \dot{e}_{pi}\right) + B_{pi}F_i^R + \Delta_{pi}. \end{aligned} \tag{2.20}$$

Let $z_{pi} = \left[e_{pi}^T \quad \dot{e}_{pi}^T \right]^T$ and $z_{\Theta i} = \left[e_{\Theta i}^T \quad \dot{e}_{\Theta i}^T \right]^T$. From (2.9), (2.16), (2.17), and (2.20), the closed-loop node error system can be written as

$$\begin{aligned} \dot{z}_{pi} = & A_p z_{pi} - \alpha_p B_z \sum_{j\in N_i} w_{ij}\left(K_p\left(e_{pi} - e_{pj}\right) + K_v\left(\dot{e}_{pi} - \dot{e}_{pj}\right)\right) \\ & - \alpha_p b_{li} B_z\left(K_p e_{pi} + K_v \dot{e}_{pi}\right) + B_z\left(B_{pi}F_i^R + \Delta_{pi}\right), \\ \dot{z}_{\Theta i} = & A_\Theta z_{\Theta i} + B_z\left(B_{\Theta i}u_{\Theta i}^R + \Delta_{\Theta i}\right),\ i \in \Phi, \end{aligned} \tag{2.21}$$

where

$$A_p = \begin{bmatrix} 0_{3\times 3} & I_3 \\ 0_{3\times 3} & 0_{3\times 3} \end{bmatrix}, A_\Theta = \begin{bmatrix} 0_{3\times 3} & I_3 \\ -K_\Theta & -K_\omega \end{bmatrix}, B_z = \begin{bmatrix} 0_{3\times 3} \\ I_3 \end{bmatrix}.$$

Then, the global error dynamics of the whole quadrotor group can be given by

$$\begin{aligned} \dot{z}_p & = \left(I_N \otimes A_p - \alpha_p (L + B_L) \otimes B_z K_z\right) z_p + \left(I_N \otimes B_z\right)\tilde{\Delta}_p \\ & = \bar{A}_{pc} z_p + \bar{B}_\Delta \tilde{\Delta}_p, \end{aligned} \tag{2.22}$$

and

$$\dot{z}_\Theta = \left(I_N \otimes A_\Theta\right) z_\Theta + \left(I_N \otimes B_z\right)\tilde{\Delta}_\Theta = \bar{A}_{\Theta c} z_\Theta + \bar{B}_\Delta \tilde{\Delta}_\Theta, \tag{2.23}$$

where $z_p = \left[z_{pi} \right] \in \mathbb{R}^{6N\times 1}$, $z_\Theta = \left[z_{\Theta i} \right] \in \mathbb{R}^{6N\times 1}$, $\tilde{\Delta}_p = \left[B_{pi}F_i^R + \Delta_{pi} \right] \in \mathbb{R}^{3N\times 1}$, $\tilde{\Delta}_\Theta = \left[B_{\Theta i}u_{\Theta i}^R + \Delta_{\Theta i} \right] \in \mathbb{R}^{3N\times 1}$, $B_L = diag\left\{ b_{li} \right\} \in \mathbb{R}^{N\times N}$, and $K_z = \left[K_p \quad K_v \right]$.

Lemma 2.1

Let the design matrices $Q_p = Q_p^T \in \mathbb{R}^{6\times 6}$ and $\Pi_p = \Pi_p^T \in \mathbb{R}^{3\times 3}$ be symmetric and positive definite. Let the design matrices K_Θ, K_ω be diagonal matrices with positive elements. Design the nominal position controller parameter matrix K_z as

$$K_z = \Pi_p^{-1} B_z^T P_p,$$

where P_p is the unique positive definite solution of the following associated Riccati equation

$$A_p^T P_p + P_p A_p + Q_p - P_p B_z \Pi_p^{-1} B_z^T P_p = 0.$$

It can be seen that all the eigenvalues of $\bar{A}_{\Theta c}$ have negative real parts, and thereby $\bar{A}_{\Theta c}$ is asymptotically stable. If the root can receive the information from the virtual leader and the graph G has a spanning tree, then $\alpha_p \geq 0.5\lambda_{pr}^{\min}$, where $\lambda_{pr}^{\min} = \min_{i\in\Phi} \mathrm{Re}(\lambda_{pi})$ and λ_{pi} indicates the eigenvalues of $(L + B_L)$ and $\bar{A}_{pc}$ is asymptotically stable.

Proof According to Theorem 1 in [52], $\bar{A}_{pc}$ is asymptotically stable. In addition, if K_Θ and K_ω are diagonal matrices with positive elements, $\bar{A}_{\Theta c}$ is also asymptotically stable.

Theorem 2.1

Consider the dynamical model of the quadrotor as depicted in (2.6), the robust formation controller consisting of (2.10), (2.11), (2.14), (2.15), (2.16), (2.17), and (2.19), and the nominal controller parameters determined by Lemma 2.1. If the graph G has a spanning tree, the root can obtain the information from the virtual leader, and the initial states $z_p(0)$ and $z_\Theta(0)$ are bounded, then for any given positive constant ε_e, there exist finite positive constants $T^*, \eta_p^*, \eta_\Theta^*$, such that for any $\eta_{pj,i} \geq \eta_p^*$ and $\eta_{\Theta j,i} \geq \eta_\Theta^*$ $(i \in \Phi)$, then one can obtain that all states involved are bounded and the position error e_{pi} and the yaw error $e_{\psi i}$ of each quadrotor satisfy that $\max_j |e_{pj,i}(t)| \leq \varepsilon_e$ and $|e_{\psi i}(t)| \leq \varepsilon_e$, $\forall t \geq T^*$, respectively.

Proof 2.1 Combining (2.12), (2.16), (2.22), and (2.23) yields that

$$\dot{z}_p = \bar{A}_{pc} z_p + \bar{B}_\Delta (I_{3N} - \Gamma_p(s)) \Delta_p,$$

$$\dot{z}_\Theta = \bar{A}_{\Theta c} z_\Theta + \bar{B}_\Delta (I_{3N} - \Gamma_\Theta(s)) \Delta_\Theta,$$

where $\Gamma_p(s) = diag\{\Gamma_{pi}(s)\} \in \mathbb{R}^{3N\times 3N}$, $\Gamma_\Theta(s) = diag\{\Gamma_{\Theta i}(s)\} \in \mathbb{R}^{3N\times 3N}$, $\Delta_p = [\Delta_{pi}] \in \mathbb{R}^{3N\times 1}$, and $\Delta_\Theta = [\Delta_{\Theta i}] \in \mathbb{R}^{3N\times 1}$. It follows that

$$\begin{aligned} \|z_p\|_\infty &\leq \pi_{z_p(0)} + \chi_{Bp} \|\Delta_p\|_\infty, \\ \|z_\Theta\|_\infty &\leq \pi_{z_\Theta(0)} + \chi_{B\Theta} \|\Delta_\Theta\|_\infty, \end{aligned} \tag{2.24}$$

and

$$\begin{aligned}
&\max_j \left|z_{pj}(t)\right| \le \max_j \left|c_{6N,j}^T e^{\bar{A}_{pc}t} z_p(0)\right| + \chi_{Bp}\left\|\Delta_p\right\|_\infty, \\
&\max_j \left|z_{\Theta j}(t)\right| \le \max_j \left|c_{6N,j}^T e^{\bar{A}_{\Theta c}t} z_\Theta(0)\right| + \chi_{B\Theta}\left\|\Delta_\Theta\right\|_\infty,
\end{aligned} \tag{2.25}$$

where

$$\pi_{z_p(0)} = \max_j \sup_{t\ge 0}\left|c_{6N,j}^T e^{\bar{A}_{pc}t} z_p(0)\right|,$$

$$\pi_{z_\Theta(0)} = \max_j \sup_{t\ge 0}\left|c_{6N,j}^T e^{\bar{A}_{\Theta c}t} z_\Theta(0)\right|,$$

$$\chi_{Bp} = \left\|(sI_{6N} - \bar{A}_{pc})^{-1}\bar{B}_\Delta(I_{3N} - \Gamma_p(s))\right\|_1,$$

$$\chi_{B\Theta} = \left\|(sI_{6N} - \bar{A}_{\Theta c})^{-1}\bar{B}_\Delta(I_{3N} - \Gamma_\Theta(s))\right\|_1.$$

From Remark 2.6, one can have that the gains of the robust filters $\Gamma_{pj,i}(s)$ and $\Gamma_{\Theta j,i}(s)$ are closer to 1 by selecting positive parameters $\eta_{pj,i}$ and $\eta_{\Theta j,i}$ $(i \in \Phi)$. In this case, χ_{Bp} and $\chi_{B\Theta}$ can be made as small as desired by selecting $\eta_{pj,i}$ and $\eta_{\Theta j,i}$ properly. Since Δ_Θ in (2.8) is norm bounded, one can have that positive constants $\varepsilon_{e\Theta}$ and $\varepsilon_{\Delta\Theta}$ exist such that

$$\left\|\Delta_\Theta\right\|_\infty \le \varepsilon_{e\Theta}\left\|z_\Theta\right\|_\infty + \varepsilon_{\Delta\Theta}. \tag{2.26}$$

If the robust filter parameters $\eta_{\Theta j,i}$ $(i \in \Phi)$ are selected such that $\chi_{B\Theta} < \varepsilon_{e\Theta}^{-1}$, one can yield the following inequality by substituting (2.26) into the second equation of (2.24) as

$$\left\|z_\Theta\right\|_\infty \le \frac{\pi_{z_\Theta(0)} + \chi_{B\Theta}\varepsilon_{\Delta\Theta}}{1 - \chi_{B\Theta}\varepsilon_{e\Theta}}. \tag{2.27}$$

For the bounded initial state $z_\Theta(0)$, one can see from (2.27) that z_Θ is bounded, and thereby Δ_Θ is bounded. Besides, one can obtain that the states of the robust compensating input $u_{\Theta i}^R$ and the composite control input $u_{\Theta i}$ are bounded. Now, it can be obtained that all states involved in the rotational dynamics are bounded.

Furthermore, there exist positive constants $\varepsilon_{ep3,i}$ and $\varepsilon_{\Delta p3,i}$ such that

$$\left\|\Delta_{p3,i}\right\|_\infty \le \varepsilon_{ep3,i}\left\|z_{p3,i}\right\|_\infty + \varepsilon_{\Delta p3,i},\ i \in \Phi, \tag{2.28}$$

where $z_{p3,i} = \begin{bmatrix} e_{p3,i} & \dot{e}_{p3,i} \end{bmatrix}^T$. Similar to the derivations of the rotational dynamics as shown in (2.27), one can obtain that all states in the vertical dynamics of the global system are bounded. Therefore, it follows that the vertical control input u_{zi} is bounded, and thereby $\tilde{F}_i$ is bounded. From (2.8), one can obtain positive constants $\varepsilon_{\Delta p1,i}$ and $\varepsilon_{\Delta p2,i}$ such that

$$\left\|\Delta_{p1,i}\right\|_{\infty} \leq \varepsilon_{\Delta p1,i},$$

$$\left\|\Delta_{p2,i}\right\|_{\infty} \leq \varepsilon_{\Delta p2,i},\ i \in \Phi.$$

Therefore, there exists a positive constant $\varepsilon_{\Delta p}$ such that

$$\left\|\Delta_p\right\|_{\infty} \leq \varepsilon_{\Delta p}. \tag{2.29}$$

From (2.24) and (2.29), one can have that

$$\left\|z_p\right\|_{\infty} \leq \pi_{z_p(0)} + \chi_{Bp}\varepsilon_{\Delta p}. \tag{2.30}$$

One can see that for the bounded initial state $z_p(0)$, z_p is also bounded. In addition, one can obtain that the states of the robust compensating input F_i^R $(i \in \Phi)$ and the composite control input F_i are bounded. It follows that all states involved in the translational dynamics are bounded.

Now, from (2.25), (2.26), and (2.29), one can have that

$$\begin{aligned}
\max_j \left|z_{pj}(t)\right| &\leq \max_j \left|c_{6N,j}^T e^{\bar{A}_{pc}t} z_p(0)\right| + \chi_{Bp}\varepsilon_{\Delta p}, \\
\max_j \left|z_{\Theta j}(t)\right| &\leq \max_j \left|c_{6N,j}^T e^{\bar{A}_{\Theta c}t} z_{\Theta}(0)\right| + \chi_{B\Theta}\varepsilon_{\Delta\Theta}.
\end{aligned} \tag{2.31}$$

Since $\bar{A}_{pc}$ and $\bar{A}_{\Theta c}$ are asymptotically stable as shown in Lemma 5.1, there exist positive constants $T^*, \eta_p^*, \eta_{\Theta}^*$, such that for any $\eta_{pj,i} \geq \eta_p^*$ and $\eta_{\Theta j,i} \geq \eta_{\Theta}^*$ $(i \in \Phi)$, then one has that $\max_j \left|e_{pj,i}(t)\right| \leq \varepsilon_e$ and $\left|e_{\psi i}(t)\right| \leq \varepsilon_e$, $\forall t \geq T^*$.□

Remark 2.9

It should be noted that the robust compensator parameters $\eta_{pj,i}$ and $\eta_{\Theta j,i}$ $(i \in \Phi)$ determined by Theorem 2.1 are conservative, which means that the theoretical value of $\eta_{pj,i}$ and $\eta_{\Theta j,i}$ $(i \in \Phi)$ may be much larger than its real one. Therefore, in practical applications, one can determine the robust compensator parameters using an online tuning method: first, set $\eta_{pj,i}$ and $\eta_{\Theta j,i}$ $(i \in \Phi)$ with initial small values; second, increase $\eta_{pj,i}$ and $\eta_{\Theta j,i}$ $(i \in \Phi)$ until the desired tracking performance of the proposed closed-loop control system is obtained.

Remark 2.10

If the second derivative of the attitude reference, e.g., $\ddot{\Theta}_{ri}$ $(i \in \Phi)$, cannot be obtained directly in practical applications, it can be ignored in the nominal attitude control law, as shown in (2.13). In this case, $\ddot{\Theta}_{ri}$ can be considered as disturbances and included in the equivalent disturbance $\Delta_{\Theta i}$. The robustness properties of the global swarm system can still be guaranteed.

2.4 Numerical Simulation Results

In order to validate the proposed control algorithms presented in this chapter, the control strategy is implemented on a team of three quadrotors, whose dynamics is subject to underactuation, nonlinearities, and disturbances. The quadrotor parameter matrices are selected as $B_{\Theta i} = diag\{9.2, 9.7, 15.99\}$ and $B_{pi} = diag\{1,1,1\}$, $m_i = 2$ kg, $g_i = 9.81$ m/s^2, and $J_i = diag\{0.109, 0.103, 0.0625\}$ kg $\cdot$ m^2. The positions of the three quadrotors are required to maintain a constant formation pattern, described by a triangle to the horizontal plane in the inertial coordinate as follows: $\delta_1 = \begin{bmatrix} 1 & 1 & 0 \end{bmatrix}^T$, $\delta_2 = \begin{bmatrix} -1 & 1 & 0 \end{bmatrix}^T$, and $\delta_3 = \begin{bmatrix} -1 & -1 & 0 \end{bmatrix}^T$, while the yaw angle of each vehicle is required to be stabilized at 0 deg. The controller parameters in the numerical simulations are selected as follows: $K_\Theta = diag\{50,50,50\}$, $K_\omega = diag\{35,35,35\}$, $K_v = diag\{0.7,0.7,0.7\}$, $K_p = diag\{1,1,1\}$, $\alpha_p = 5$, $\eta_{\Theta j,i} = 20$, and $\eta_{pj,i} = 1$. Set $\varepsilon_e = 0.1$, and a simulation test is conducted to check whether the tracking errors converge into the neighborhood of the origin bounded by ε_e in a finite time, as shown in Theorem 2.1. The formation mission is implemented by the group of quadrotors with bounded initial states. The initial conditions of the four quadrotors are selected as $p_1(0) = \begin{bmatrix} 1.5 & 1.5 & -1 \end{bmatrix}^T$, $p_2(0) = \begin{bmatrix} -1 & 2 & -1 \end{bmatrix}^T$, $p_3(0) = \begin{bmatrix} -1 & -2 & -1 \end{bmatrix}^T$, $\dot{p}_1(0) = \begin{bmatrix} -0.1 & 0.1 & 0.1 \end{bmatrix}^T$, $\dot{p}_2(0) = \begin{bmatrix} 0.2 & -0.1 & -0.1 \end{bmatrix}^T$, $\dot{p}_3(0) = \begin{bmatrix} 0.1 & -0.2 & 0.1 \end{bmatrix}^T$, $\Theta_i(0) = 0_{3\times1}$, and $\dot{\Theta}_i(0) = 0_{3\times1}$.

In the simulation, there are five vehicles and the communication topology of vehicles is described in Figure 2.1. The set of edges $E = \{(v_2, v_1), (v_3, v_2), (v_4, v_1)\}$, and the weighted adjacency matrix $W = \begin{bmatrix} w_{ij} \end{bmatrix}$ with $w_{ij} = 0.5$ for $(v_i, v_j) \in E$ and 0 otherwise. The root v_1 can obtain the information from the virtual leader, and $b_{li} = 0.5$ for $i = 1$ and 0 otherwise. The reference trajectory of the virtual leader is described by $p_{r0}(t) = \begin{bmatrix} 0.4t & 0.4t & 0.8t \end{bmatrix}^T$. The ith quadrotor dynamics is subject to the nonvanishing and time-varying external atmospheric forces and torques as $(-1)^i \begin{bmatrix} 0.1\cos(t) & 0.1\sin(t) & 0.1\cos(t) \end{bmatrix}^T$ and $(-1)^i \begin{bmatrix} \sin(t) & \cos(t) & \sin(t) \end{bmatrix}^T$, respectively.

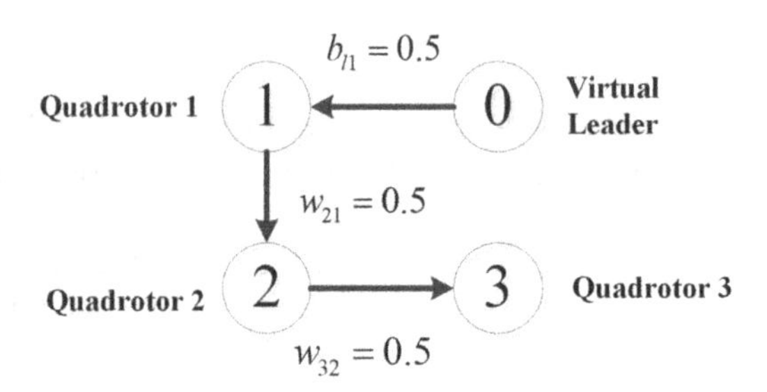

FIGURE 2.1
Desired formation pattern and communication graph.

The three-dimensional (3-D) trajectories of the three quadrotors are depicted in Figure 2.2. The longitudinal, lateral, and vertical positions are depicted in Figure 2.3. The translational velocities, the Euler angles, and the position errors are given in Figures 2.4, 2.5, and 2.6, respectively. The dashed lines represent the formation patterns. From these figures, it can be observed that the position tracking errors can converge into the neighborhood of the origin bounded. The proposed global closed-loop control system can achieve good tracking performances and robustness stability properties. The effects of nonlinear dynamics, parametric perturbations, and external disturbances can be restrained by the proposed formation protocol.

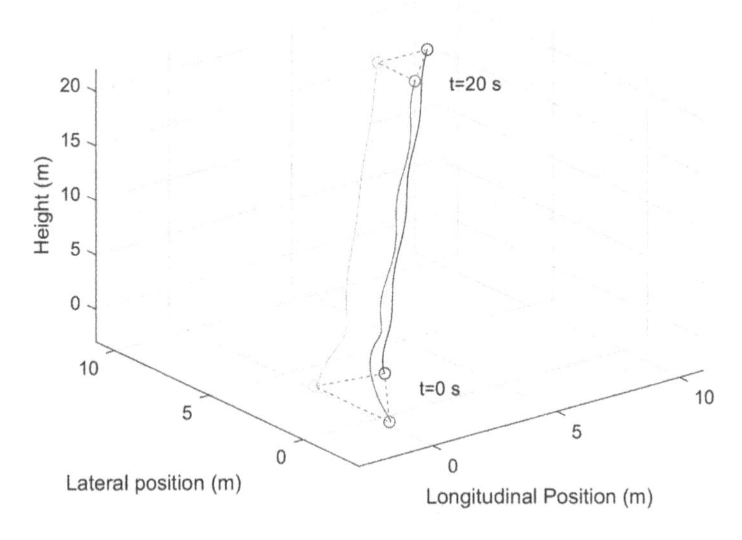

FIGURE 2.2
Three-dimensional trajectories of three quadrotors.

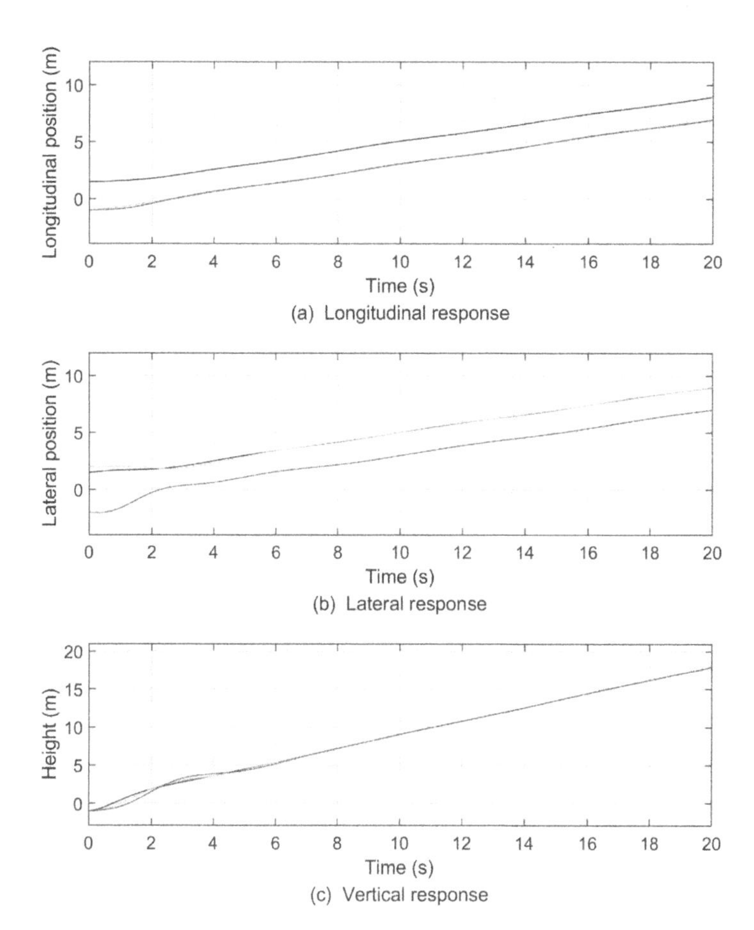

FIGURE 2.3
Position responses of three quadrotors.

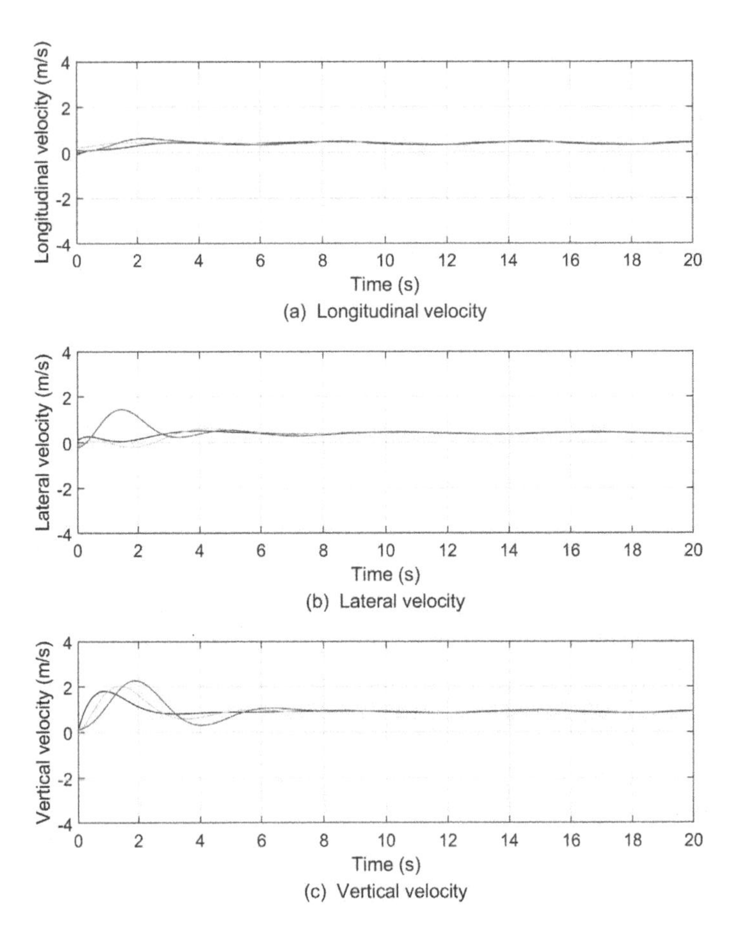

FIGURE 2.4
Translational velocity responses of three quadrotors.

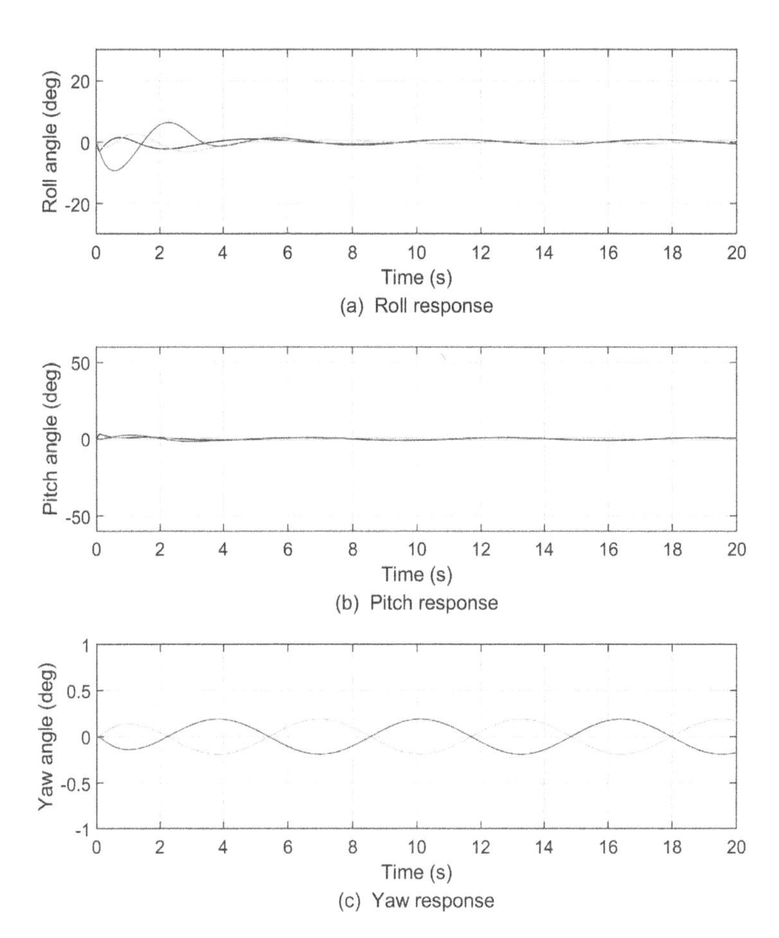

FIGURE 2.5
Euler angle responses of three quadrotors.

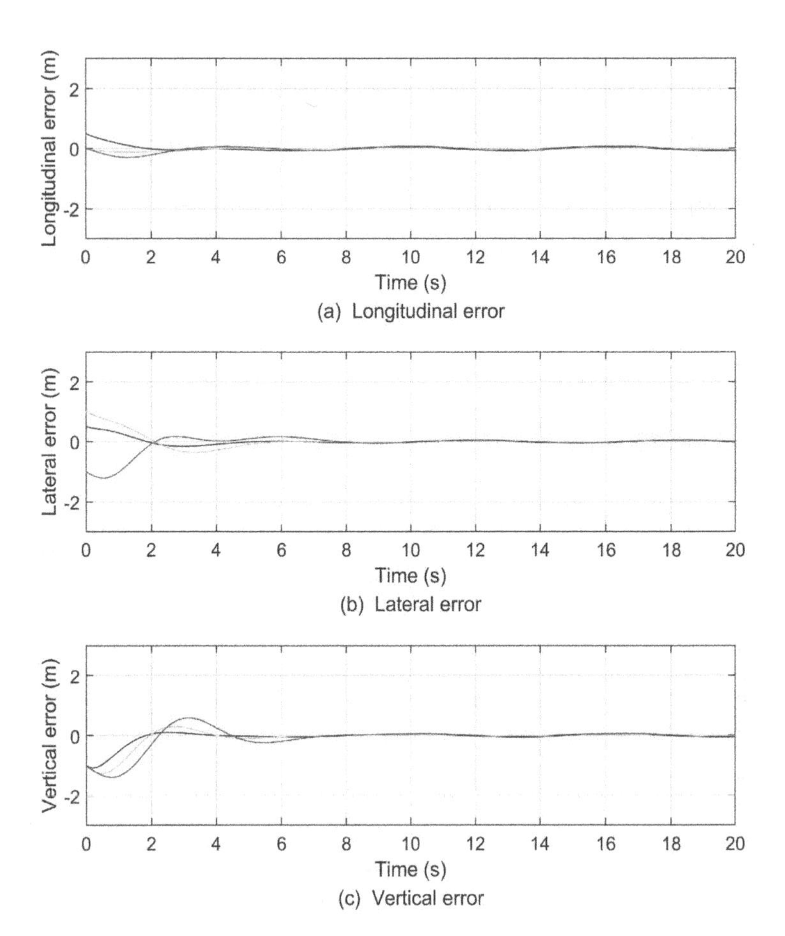

FIGURE 2.6
Position errors of three quadrotors.

2.5 Conclusion

The formation control problem of a team of quadrotors with underactuated, highly nonlinear and strongly coupled dynamics, and disturbances is addressed in this chapter. A distributed controller is proposed based on the robust compensation theory and the backstepping approach. The proposed robust formation controller includes a position controller to achieve the desired formation trajectories and patterns and an attitude controller to regulate the attitudes of quadrotors. The robustness properties of the global closed-loop control system are proven. Numerical simulation studies are provided to show the effectiveness and the advantages of the proposed robust formation protocol.

3

Robust Formation Trajectory Tracking Control for Multiple Quadrotors with Communication Delays

Communication delays are inherently present in information exchange between quadrotors and have an effect on the control performance of quadrotor formation. This chapter investigates the robust formation control problem for a team of quadrotors with communication delays. The dynamics of each quadrotor is subject to nonlinearities, parametric perturbations, and external disturbances. A distributed formation controller is developed for the quadrotor team to overcome the unknown delays. The designed controller consists of a position controller to achieve the desired formation trajectories and patterns and an attitude controller to regulate the attitude angles for each quadrotor. The robustness of the constructed global closed-loop control system can be proven by the Lyapunov theorem. Experimental results on multiple micro quadrotors are provided to verify the effectiveness of the proposed control method.

3.1 Introduction

In recent years, formation flight control has been extensively studied in the literatures for a broad range of applications in various areas, such as communication relay, search and rescue, and persistent reconnaissance, as illustrated in [54–57]. Multiple unmanned vehicles (UAVs) cooperate with each other to solve some tasks which are difficult for the traditional single aircraft. Over the past two decades, several classical formation control methods have been developed to achieve formation control for multi-agent systems, such as the leader-follower method, the behavioral strategy, and the virtual structure approach (see, *e.g.*, [14, 16, 19, 26, 38, 44]). However, leader-follower, behavior, and virtual-structure–based formation control approaches have their own strengths and weaknesses. Recently, a consensus-based approach has been developed, which serves as a promising solution to address the formation control problem of a team of UAVs, as shown in [11, 21, 22, 23, 27, 28, 60, 61]. For example, leader-follower approaches are easy to implement but have no

DOI: 10.1201/9781003242147-3

explicit feedback on the formation and lack of robustness due to the existence of the leader, behavior approaches can ensure collision avoidance but are difficult to analyze mathematically, and so on.

Quadrotor UAVs served as popular aerial platforms possessing various advantages such as low-cost, simple structure, vertical take-off and landing, as illustrated in [51, 59]. In prior works, several typical approaches are provided for a team of UAVs to achieve formation flight missions with free time delay. In [38], an integrated optimal control framework was developed for fixed-wing UAVs to achieve formation trajectory tracking. In [19], a potential field-theory–based control scheme was proposed to achieve formation flight. In [55, 57], the formation control for a set of UAVs was addressed. In [44], a hybrid three-dimensional formation control framework was presented under the leader-follower context using UAVs with simplified translational dynamics. In [11], the dynamical model of each vehicle was considered as a second-order linear system and the formation experiments were implemented. In [19] and [38], the formation control problem for multiple nonlinear vehicles was discussed. However, the vehicle dynamics was simplified and the formation control problems for such underactuated systems were not fully studied in [11, 44, 55, 57].

Furthermore, the formation control problems for a team of uncertain systems were studied in [16, 22, 23, 25]. In [16], the formation control problem for a team of uncertain quadrotors was discussed based on the classical leader-follower method, and a H_∞ control controller was developed to restrain the effects of model uncertainties and external disturbances. In [22] and [23], a disturbance estimator was constructed for multiple 3-degrees-of-freedom (3-DOF) helicopters in the presence of model uncertainties and external disturbances in the rotational dynamics, but the underactuated 6-DOF aerial vehicle system was not further discussed. In [25], a robust formation control strategy was proposed for a team of nonlinear systems with parametric perturbations.

Noteworthy, time delays in the existing communication mechanism are inevitable in practical applications. It is worthwhile to point out that the above-mentioned results with a straight application to the case with communication delay will deteriorate the control performance. Moreover, several works have been conducted to address the robust formation control problem for a team of UAVs subject to communication delays.

In [26], a model predictive formation control method was proposed to achieve the formation flight control for multiple quadrotors with communication delays. In [27], a sufficient condition was proposed for the consensus of discrete-time multi-agent systems composed of double-order dynamics with bounded communication delays. In [28], a formation control method was developed for the second-order agents to achieve the desired formation trajectory tracking, and a neighbor-based feedback control rule was designed to reduce the effects of communication delays between agents. However, the effects of other uncertainties were not fully discussed in the stability property analysis of the constructed closed-loop control systems in [26–28].

In this chapter, one focuses on solving the formation flight problem of UAVs in the presence of communication delay. A robust formation controller is proposed for a team of micro quadrotors subject to communication delays. The proposed formation controller consists of a position controller to achieve the desired formation trajectories and patterns and an attitude controller to regulate the attitude dynamics of the quadrotors. Both position controller and the attitude controller are developed based on the linear quadratic regulator (LQR) control method and the robust compensation theory.

Compared to existing literatures, the main contributions of this chapter can be summarized as follows: First, the model of each quadrotor is an *underactuated nonlinear* system involving *three translational DOF* and *three rotational DOF* and the distributed indoor flight experiments for multiple micro quadrotors swarm systems to achieve time-varying formation are presented to demonstrate the advantages of the proposed formation control protocol. However, the formation control problems for such underactuated 6-DOF systems with completely nonlinear and coupled terms were not further discussed in [11, 19, 22, 23, 44, 58]. Second, the influence of the *parametric perturbations* and *external disturbances* are considered in the dynamical model. However, in [11, 21, 27, 28, 44, 47], the uncertainty rejection problems were not fully considered in the robust stability analysis of the constructed closed-loop control systems. Third, the influence of the *communication delays* can be restrained by the proposed robust formation control protocol, and the tracking errors can converge into a neighborhood of the origin bounded by a given constant in a finite time. However, the effects of the communication delays on the global closed-loop control systems were not further studied in [14, 16, 19, 21–24, 44, 47].

3.2 Preliminaries and Problem Formulation

3.2.1 Quadrotor Model

Consider N quadrotors modeled as rigid bodies. Let I denote the earth-fixed inertial frame and $\mathcal{B}$ the body-fixed frame attached to quadrotor i with the origin in its mass center. Define $p_i = \left[p_{xi}, p_{yi}, p_{zi}\right]^T \in \mathbb{R}^{3\times1}$ and $v_i = \left[v_{xi}, v_{yi}, v_{zi}\right]^T \in \mathbb{R}^{3\times1}$ as the position and the linear velocity of quadrotor i expressed in the inertial frame.

As depicted in [51], the translational dynamics and rotational dynamics of quadrotor i can be written as

$$
\begin{aligned}
m_i \ddot{p}_i &= R_{bi} f_{bi} + d_{pi}, \\
J_i \ddot{\Theta}_i &= \tau_{bi} + C_i(\Theta_i, \dot{\Theta}_i)\dot{\Theta}_i + d_{\Theta i},
\end{aligned}
\tag{3.1}
$$

where m_i and J_i represent the mass and the inertia matrix of quadrotor i, respectively, $C_i(\Theta_i, \dot{\Theta}_i)$ represents the Coriolis term as illustrated in [51], d_{pi} and $d_{\Theta i}$ are the external disturbances, and $\Theta_i = [\phi_i,\ \theta_i,\ \psi_i]^T \in \mathbb{R}^{3\times 1}$ indicates the three Euler angles, i.e., the roll, pitch, and yaw angles, respectively. In order to avoid the singularity problem, the attitude of quadrotor i is assumed to satisfy: $|\phi_i| < \pi/2$, $|\theta_i| < \pi/2$, and $|\psi_i| < \pi/2$, successively. The rotation matrix $R_{bi} \in SO(3)$ from the body-fixed coordinate to the earth-fixed inertial coordinate can be expressed as

$$
R_{bi} = \begin{bmatrix}
\cos\theta_i \cos\psi_i & \cos\psi_i \sin\phi_i \sin\theta_i - \cos\phi_i \sin\psi_i & \sin\phi_i \sin\psi_i + \cos\phi_i \cos\psi_i \sin\theta_i \\
\cos\theta_i \sin\psi_i & \cos\phi_i \cos\psi_i + \sin\phi_i \sin\theta_i \sin\psi_i & \cos\phi_i \sin\theta_i \sin\psi_i - \cos\psi_i \sin\phi_i \\
-\sin\theta_i & \cos\theta_i \sin\phi_i & \cos\phi_i \cos\theta_i
\end{bmatrix}.
$$

The external force f_{bi} and torque τ_{bi} relative to the body-fixed coordinate of the quadrotor can be described as

$$
f_{bi} = \begin{bmatrix} 0 \\ 0 \\ T_f \end{bmatrix} + R_{bi}^T \begin{bmatrix} 0 \\ 0 \\ -m_i g \end{bmatrix}, \tau_{bi} = \begin{bmatrix} l_{bi} k_{\sigma i}(\sigma_{2,i}^2 - \sigma_{4,i}^2) \\ l_{bi} k_{\sigma i}(\sigma_{1,i}^2 - \sigma_{3,i}^2) \\ k_{\tau i} \sum_{r=1}^{4} (-1)^{r+1} \sigma_{r,i}^2 \end{bmatrix},
$$

where g denotes the gravity constant, l_{bi}, $k_{\sigma i}$, and $k_{\tau i}$ are positive scale factors, and $\sigma_{r,i}$ $(r = 1,2,3,4)$ denote the rotational velocity of rotor r. The total lift $T_f = k_{\sigma i} \sum_{r=1}^{4} \sigma_{r,i}^2$ is assumed to be positive because $\sigma_{r,i}$ is nonnegative and $T_f = 0$ results in a state of free for quadrotor i. The control input commands for quadrotor i can be denoted as follows:

$$
u_{zi} = \sum_{r=1}^{4} \sigma_{r,i}^2,\ u_{\Theta 1,i} = \sigma_{2,i}^2 - \sigma_{4,i}^2,\ u_{\Theta 2,i} = \sigma_{1,i}^2 - \sigma_{3,i}^2,\ u_{\Theta 3,i} = \sum_{r=1}^{4} (-1)^{r+1} \sigma_{r,i}^2.
$$

Remark 3.1

It should be noted that the quadrotor model consists of 6-DOFs (three translational DOFs and three rotational DOFs) but four control inputs. Therefore, the quadrotor system is underactuated. In this chapter, the position and heading control mode is chosen for each quadrotor, including the longitudinal position p_x, the lateral position p_y, the height p_z, and the yaw angle ψ as outputs, and u_θ, u_ϕ, u_z, and u_ψ as control inputs.

3.2.2 Problem Formulation

Define $B_{pi} = m_i^{-1} I_3 k_{\sigma i}$ and $B_{\Theta i} = J_i^{-1} diag\{l_b k_{\sigma i},\ l_b k_{\sigma i},\ k_{\tau i}\}$ $(i \in \Pi)$. The parameters can be divided into two parts: the nominal part represented by the superscript N and the uncertain part represented by the superscript Δ, and, for instance, $B_{\Theta i} = B_{\Theta i}^N + B_{\Theta i}^\Delta$. Let $B_p^N = diag(b_{p1}^N, b_{p2}^N, b_{p3}^N)$ and $B_\Theta^N = diag(b_{\Theta 1}^N, b_{\Theta 2}^N, b_{\Theta 3}^N)$. Then, for quadrotor i, from (3.1), one can obtain that

$$
\begin{aligned}
\ddot{p}_i &= B_{pi}^N u_{pi} - g c_{3,3} + \Delta_{pi}, \\
\ddot{\Theta}_i &= B_{\Theta i}^N u_{\Theta i} + (J_i^N)^{-1} C(\Theta_i, \dot{\Theta}_i)\dot{\Theta}_i + \Delta_{\Theta i},
\end{aligned}
\tag{3.2}
$$

where $u_{\Theta i} = \left[u_{\Theta 1,i},\ u_{\Theta 2,i},\ u_{\Theta 3,i}\right]^T$. $u_{pi} = \left[u_{p1,i},\ u_{p2,i},\ u_{p3,i}\right]^T$ is the virtual position control input and satisfies

$$
u_{pi} = u_{zi} \begin{bmatrix} \sin\phi_i \sin\psi_i + \cos\phi_i \cos\psi_i \sin\theta_i^r \\ \cos\phi_i \sin\theta_i \sin\psi_i - \cos\psi_i \sin\phi_i^r \\ \cos\phi_i \cos\theta_i \end{bmatrix}. \tag{3.3}
$$

Δ_{pi} and $\Delta_{\Theta i}$ are named equivalent disturbances, which include the parametric perturbations and external disturbances, and can be given by

$$
\begin{aligned}
\Delta_{pi} &= B_{pi}^N \tilde{u}_{pi} + m_i^{-1} d_{pi}, \\
\Delta_{\Theta i} &= -(J_i^N)^{-1} C(\Theta_i, \dot{\Theta}_i)\dot{\Theta}_i + (J_i)^{-1} C(\Theta_i, \dot{\Theta}_i)\dot{\Theta}_i + B_{\Theta i}^\Delta u_{\Theta i} + J_i^{-1} d_{\Theta i},
\end{aligned}
\tag{3.4}
$$

where $\tilde{u}_{pi}$ is the force error and satisfies

$$
\tilde{u}_{pi} = \left[\tilde{u}_{p1,i}, \tilde{u}_{p2,i}, \tilde{u}_{p3,i}\right]^T = u_{zi}(B_{pi}^N)^{-1} B_{pi} R_{bi} c_{3,3} - u_{pi}.
$$

The objective of this chapter is to design a distributed formation control protocol for each quadrotor in the group such that the formation center of the quadrotors tracks a prescribed stationary trajectory and the quadrotors keep a geometrical configuration.

Let $p_r \in \mathbb{R}^{3\times 1}$ be the desired trajectory of the formation center, which can also be considered as the virtual team leader. p_r is assumed to be differentiable, satisfying $\dot{p}_0^r = v_0^r$ and $\ddot{p}_0^r = 0_{3\times 1}$. Define $\delta_{ij} = \left[\delta_{1,ij},\ \delta_{2,ij},\ \delta_{3,ij}\right]^T \in \mathbb{R}^{3\times 1}$ $(i, j \in \Pi)$ as the desired position deviation between quadrotor i and quadrotor j, and δ_{ij} determines the formation pattern of the quadrotor team.

Let $\delta_{ij} = \delta_i - \delta_j$ $(i, j \in \Pi)$, where δ_i indicates the desired position deviation between the virtual team leader and quadrotor i.

Let θ_i^r represent the desired pitch angle and ϕ_i^r the roll angle references. The yaw angle needs to track the reference ψ_i^r. Let $\Theta_i^r = \left[\phi_i^r, \theta_i^r, \psi_i^r\right]^T \in \mathbb{R}^{3\times 1}$. Then, one can obtain the position and attitude errors as $e_{pi} = p_i - \delta_i - p_r = \left[e_{pk,i}\right] \in \mathbb{R}^{3\times 1}$, $e_{vi} = \dot{e}_{pi} = \left[e_{vk,i}\right]^T \in \mathbb{R}^{3\times 1}$, $e_{\Theta i} = \Theta_i - \Theta_i^r = \left[e_{\Theta ki}\right]^T \in \mathbb{R}^{3\times 1}$, and $e_{\omega i} = \dot{e}_{\Theta i} = \left[e_{\omega k,i}\right]^T \in \mathbb{R}^{3\times 1}$ $(i \in \Pi)$.

Remark 3.2

Model (3.2) represents the real vehicle model. By ignoring the equivalent disturbances Δ_{pi} and $\Delta_{\Theta i}$ $(i \in \Phi)$, the vehicle model (3.2) represents the nominal model. The real model can be considered as the nominal model added by the equivalent disturbances.

3.3 Controller Design

In this section, the formation protocol design problem for the group of quadrotors is investigated. The formation controller is designed based on the Linear Quadratic Regulation (LQR) approach and the robust compensation theory as illustrated in [62]. For each quadrotor, the proposed robust controller consists of a trajectory tracking controller to track the desired formation trajectory and form the desired geometrical configurations and an attitude controller to align the attitude angles in the rotational motion.

3.3.1 Position Controller Design

The virtual position control input $u_{pi}(t)$ consists of two parts:

$$u_{pi}(t) = u_{pi}^N(t) + u_{pi}^R(t), \; i \in \Pi, \tag{3.5}$$

where $u_{pi}^N(t) \in \mathbb{R}^{3\times 1}$ is the nominal control input to achieve the desired tracking performance for the nominal translational system and $u_{pi}^R(t) \in \mathbb{R}^{3\times 1}$ the robust compensating input to restrain the influences of equivalent disturbance on the real translational system.

When quadrotor i receives the information (position and velocity) from its neighbor, i.e., quadrotor j, there exist communication delays in the information. Let ρ_1 represent the communication delays between quadrotor i and quadrotor j or the virtual team leader. The communication delays are assumed to be nonnegative and piecewise continuous. The nominal control input u_{pi}^N can be designed as follows:

$$\begin{aligned} u_{pi}^N(t) = &-\mu_P \sum_{j \in N_i} \hat{w}_{ij} \left(B_{pi}^N\right)^{-1} K_p \left(p_i(t) - p_j(t-\rho_1) - \pi_{ij} + \dot{p}_i(t) - \dot{p}_j(t-\rho_1)\right) \\ &- \mu_P \sum_{j \in N_i} \hat{w}_{ij} \left(B_{pi}^N\right)^{-1} K_v \left(\dot{p}_i(t) - \dot{p}_j(t-\rho_1)\right) \\ &- \mu_P \alpha_{pi} \left(B_{pi}^N\right)^{-1} K_p \left(p_i(t) - \pi_i - p_r(t-\rho_1)\right) \\ &- \mu_P \alpha_{pi} \left(B_{pi}^N\right)^{-1} K_v \left(\dot{p}_i(t) - \dot{p}_r(t-\rho_1)\right) + g \left(B_{pi}^N\right)^{-1} c_{3,3}, \end{aligned} \tag{3.6}$$

where μ_p is a positive scalar coupling parameter and $K_p \in \mathbb{R}^{3\times3}$ and $K_v \in \mathbb{R}^{3\times3}$ are the diagonal nominal controller gain matrices. The connection weight between the virtual leader and the quadrotor i is defined as a constant α_{pi}. $\alpha_{pi} > 0$ represents that the virtual team leader can send information to quadrotor i, and $\alpha_{pi} = 0$ otherwise. The nominal trajectory tracking controller is designed by the LQR feedback approach. Define $K_L = \left[K_p, K_v\right]$ as the LQR position controller gain. Let

$$A_p = \begin{bmatrix} 0_{3\times3} & I_3 \\ 0_{3\times3} & 0_{3\times3} \end{bmatrix}, \ B_Z = \begin{bmatrix} 0_{3\times3} \\ I_3 \end{bmatrix}.$$

Consider the design matrices $Q_L = Q_L^T \in \mathbb{R}^{6\times6}$ and $\Gamma_L = \Gamma_L^T \in \mathbb{R}^{3\times3}$. As depicted in [52], the LQR position controller gain can be given by $K_L = \Gamma_L^{-1} B_Z^T P_L$, where P_L is the positive definite solution to the following Riccati equation as

$$A_p^T P_L + P_L A_p + Q_L - P_L B_L \Gamma_L^{-1} B_Z^T P_L = 0.$$

Let Δ'_{ki} $(k = p, \Theta)$ be the equivalent disturbances involving communication delays, which satisfy

$$\begin{aligned} \Delta'_{pi}(t) &= \Delta_{pi}(t) + \Delta_{puj}(t-\rho_1) - \Delta_{puj}(t), \\ \Delta'_{\Theta i}(t) &= \Delta_{\Theta i}(t), \end{aligned}$$

where

$$\Delta_{puj}(t) = \mu_P \sum_{j \in N_i} \hat{w}_{ij} \left(K_p p_j(t) + K_v \dot{p}_j(t)\right) + \mu_P \alpha_{pi} \left(K_p p_r(t) + K_v \dot{p}_r(t)\right).$$

The term $\Delta_{puj}(t-\rho_1) - \Delta_{puj}(t)$ represents the mismatch resulting from the communication delays.

Remark 3.3

It should be pointed out that ρ_1 represents the communication delays from quadrotor i to its neighbor quadrotor j, and the delays can be time varying. It can be seen that the communication delay is involved in the position control input, and thereby the global control system.

Furthermore, from (3.6), if the ideal robust compensating input can be set as $u_{pi}^R(s) = -\left(B_{fi}^N\right)^{-1} \Delta'_{pi}(s)$, the effects of $\Delta'_{pi}(s)$ on the real model can be counteracted completely. However, it is difficult to obtain $\Delta'_{pi}(s)$ from direct measurements. In this case, robust filters $F_{pi}(s) = diag(F_{p1,i}(s),\ F_{p2,i}(s),\ F_{p3,i}(s))$ are introduced, where $F_{pl,i}(s) = f_{pl,i}^2/(s + f_{pl,i})$ $(l = 1,2,3)$ are robust position filters, where $f_{pl,i} > 0$ represents the robust position filter parameter. It can be observed that if $f_{pl,i}$ $(l = 1,2,3)$ are selected with larger positive values, the frequency bandwidth of the filters $F_{pi}(s)$ would get wider and the gains of the filters would be closer to 1. The influence of the equivalent disturbance $\Delta'_{pi}(s)$ can be restrained. Therefore, we can design the robust compensating input based on the robust filter as follows:

$$u_{pi}^R(s) = -\left(B_{fi}^N\right)^{-1} F_{pi}(s)\Delta'_{pi}(s), \tag{3.7}$$

However, it is not easy to obtain the equivalent disturbance $\Delta'_{pi}(s)$ directly in practical applications. From (3.2), one can obtain that

$$\Delta'_{pi}(t) = \ddot{p}_i(t) - B_{pi}^N u_{pi}(t) + gc_{3,3}. \tag{3.8}$$

Substituting (3.8) into (3.7), one can obtain that

$$\dot{z}_{1i}^p(t) = -f_{pi}z_{1i}^p(t) - f_{pi}^2 p_i(t) + B_{pi}^N u_{pi}(t) - gc_{3,3},$$

$$\dot{z}_{2i}^p(t) = -f_{pi}z_{2i}^p(t) + 2f_{pi}p_i(t) + z_{1i}^p(t),$$

$$u_{pi}^R(t) = \left(B_{pi}^N\right)^{-1} f_{pi}^2(z_{2i}^p(t) - p_i),$$

where $z_{1i}^p(t),\ z_{2i}^p(t) \in \mathbb{R}^{3\times1}$ are the states of the robust position filters and $f_{pi} = diag\left\{f_{pl,i}\right\} \in \mathbb{R}^{3\times3} (i \in \Pi)$.

After $u_{pi}(i \in \Pi)$ is determined by (3.3), the control input u_{zi}, the pitch angle reference $\theta_i^r(t)$, and the roll angle reference $\phi_i^r(t)$ can be expressed as follows:

$$u_{zi}(t) = u_{p3i}(t) / \cos\phi_i(t) / \cos\theta_i(t),$$

$$\theta_i^r(t) = \sin^{-1}\left((u_{p1i}(t) / u_{zi}(t) - \sin\phi_i(t)\sin\psi_i(t)) / \cos\phi_i(t) / \cos\psi_i(t))\right),$$

$$\phi_i^r(t) = \sin^{-1}\left((\cos\phi_i(t)\sin\theta_i(t)\sin\psi_i(t) - u_{p2i}(t) / u_{zi}(t)) / \cos\psi_i(t)\right).$$

3.3.2 Attitude Controller Design

Similarly to the virtual position controller design, the attitude control input $u_{\Theta i}(t)$ can be designed with the nominal control part $u_{\Theta i}^N(t)$ and the robust compensating part $u_{\Theta i}^R(t)$ as follows:

$$u_{\Theta i}(t) = u_{\Theta i}^N(t) + u_{\Theta i}^R(t), \; i \in \Pi. \tag{3.9}$$

The nominal control input $u_{\Theta i}^N(t)$ is constructed as

$$u_{\Theta i}^N(t) = \left(B_{\Theta i}^N\right)^{-1}\left((-K_\Theta e_{\Theta i}(t) - K_\omega \dot{e}_{\Theta i}(t)) + \left(J_i^N\right)^{-1} C(\Theta_i(t), \dot{\Theta}_i(t))\dot{\Theta}_i(t) + \ddot{\Theta}_{ri}(t)\right), \tag{3.10}$$

where $K_\Theta,\, K_\omega \in \mathbb{R}^{3\times 3}$ are nominal controller parameter matrices. Denote $K_R = [K_\Theta,\, K_\omega]$ as the LQR attitude controller gain. Define

$$A_R = \begin{bmatrix} 0_{3\times 3} & I_3 \\ 0_{3\times 3} & 0_{3\times 3} \end{bmatrix}.$$

Define $Q_R = Q_R^T \in \mathbb{R}^{6\times 6}$ and $\Gamma_R = \Gamma_R^T \in \mathbb{R}^{3\times 3}$ as positive definite and symmetric matrices. The LQR attitude controller gain K_R can be obtained by $K_R = \Gamma_R^{-1} B_z^T P_R$, where P_R is the positive definite solution to the following Riccati equation:

$$A_R^T P_R + P_R A_R + Q_R - P_R B_z \Gamma_R^{-1} B_z^T P_R = 0.$$

Similarly to $u_{pi}^R(s)$, $u_{\Theta i}^R(s)$ can be constructed as

$$u_{\Theta i}^R(s) = -\left(B_{\Theta i}^N\right)^{-1} F_{\Theta i}(s)\Delta'_{\Theta i}(s),$$

where $F_{\Theta l,i}(s) = diag\{f_{\Theta l,i}^2 / (s + f_{\Theta l,i})^2\} \in \mathbb{R}^{3\times 3}$ and $F_{\Theta i}(s) = diag\{F_{\Theta l,i}(s)\} \in \mathbb{R}^{3\times 3}$ with positive parameter $f_{\Theta l,i}$ to be determined. Let $f_{\Theta i} = diag\left\{f_{\Theta l,i}\right\} \in \mathbb{R}^{3\times 3}$. Then, one can obtain the realization of $u_{\Theta i}^R(t)$ in a similar way.

Remark 3.4

It can be seen that the feedback information of the controllers only depends on UAV's neighbors and itself, thus the proposed position controller and attitude controller of each quadrotor are distributed.

3.4 Robustness Property Analysis

Combining (3.2), (3.5), (3.6), (3.9), and (3.10), one can obtain that

$$\begin{aligned}\dot{\bar{X}}_{pi}(t) = A_p\bar{X}_{pi}(t) - \mu_F B_Z \sum_{j\in N_i} \hat{w}_{ij} K_p\left(e_{pi}(t) - e_{pj}(t)\right) \\ - \mu_F B_Z \sum_{j\in N_i} \hat{w}_{ij} K_v\left(e_{vi}(t) - e_{vj}(t)\right) \\ - \mu_F \alpha_{pi} B_Z\left(K_p e_{pi}(t) + K_v e_{vi}(t)\right) \\ + B_Z\left(B_{pi}^N u_{pi}^R(t) + \Delta'_{pi}(t)\right), \\ \dot{\bar{X}}_{\Theta i}(t) = A_\Theta \bar{X}_{\Theta i}(t) + B_Z\left(B_{\Theta i}^N u_{\Theta i}^R(t) + \Delta'_{\Theta i}(t)\right),\end{aligned}$$

where

$$\begin{aligned}\bar{X}_{pi}(t) = \left[e_{pi}^T(t),\ e_{vi}^T(t)\right]^T = \left[x_{pk,i}(t)\right] \in \mathbb{R}^{6\times 1}, \\ \bar{X}_{\Theta i}(t) = \left[e_{\Theta i}^T(t),\ e_{\omega i}^T(t)\right]^T = \left[x_{\Theta k,i}(t)\right] \in \mathbb{R}^{6\times 1}, i \in \Pi,\end{aligned}$$

and

$$A_\Theta = \begin{bmatrix} 0_{3\times 3} & I_3 \\ -K_\Theta & -K_\omega \end{bmatrix}.$$

Now, one can obtain the following overall closed-loop error system as

$$\begin{aligned}\dot{\bar{X}}_p(t) = A_{pc}\bar{X}_p(t) + B_{p\Delta}\tilde{\Delta}_p(t), \\ \dot{\bar{X}}_\Theta(t) = A_{\Theta c}\bar{X}_\Theta(t) + B_{\Theta\Delta}\tilde{\Delta}_\Theta(t),\end{aligned} \tag{3.11}$$

where $\tilde{\Delta}_p = \left[B_{pi}^N u_{pi}^R(t) + \Delta'_{pi}(t)\right] \in \mathbb{R}^{3N\times 1}$, $\tilde{\Delta}_\Theta = \left[B_{\Theta i}^N u_{\Theta i}^R(t) + \Delta'_{\Theta i}(t)\right] \in \mathbb{R}^{3N\times 1}$, $\bar{X}_p(t) = \left[\bar{X}_{pi}(t)\right] \in \mathbb{R}^{6N\times 1}$, $\bar{X}_\Theta(t) = \left[\bar{X}_{\Theta i}(t)\right] \in \mathbb{R}^{6N\times 1}$, $B_L = diag\left\{\alpha_{pi}\right\} \in \mathbb{R}^{N\times N}$, and

$$\begin{aligned}A_{pc} = I_N \otimes A_p - \mu_F(\hat{L} + B_L) \otimes B_Z K_L, \\ A_{\Theta c} = I_N \otimes A_\Theta, B_{p\Delta} = B_{\Theta\Delta} = I_N \otimes B_Z.\end{aligned}$$

As described in Theorem 1 of [52], if the couple gain satisfy $\mu_F \geq \vartheta_{pr}^{\min} / 2$, then $\hat{G}$ has a spanning tree and the information can flow from the virtual leader to the root, where $\vartheta_{pr}^{\min} = \min_{i \in \Pi} \mathrm{Re}(\vartheta_{pi})$ represents the minimum eigenvalues of $(\hat{L} + B_L)$. In this case, A_{pc} and $A_{\Theta c}$ are asymptotically stable matrices.

The robust compensating inputs $u_{pi}^R(t)$ and $u_{\Theta i}^R(t)$ are given in the state-space forms as

$$\begin{aligned} \dot{X}_{Rkl,i}(t) &= A_{Rkl,i}(f_{kl,i}) X_{Rkl,i}(t) + c_{2,1} \left(b_{kl,i}^N\right)^{-1} \Delta'_{kl,i}, \\ u_{kl,i}^R(t) &= -c_{2,2}^T f_{kl,i} X_{Rkl,i}(t), k = p, \Theta; l = 1,2,3, \end{aligned} \tag{3.12}$$

where

$$A_{Rkl,i}(f_{kl,i}) = \begin{bmatrix} -f_{kl,i} & 0 \\ f_{kl,i} & -f_{kl,i} \end{bmatrix}.$$

Let

$$X_{Rp}(t) = \left[X_{Rp1,i}^T(t),\ X_{Rp2,i}^T(t),\ X_{Rp3,i}^T(t)\right]^T \in \mathbb{R}^{N \times 1},$$

$$X_{R\Theta}(t) = \left[X_{R\Theta 1,i}^T(t),\ X_{R\Theta 2,i}^T(t),\ X_{R\Theta 3,i}^T(t)\right]^T \in \mathbb{R}^{N \times 1}.$$

Define $X_k(t) = \left[\bar{X}_k^T(t),\ X_{Rk}^T(t)\right]^T (k = p, \Theta)$. Let $B_{k\Delta 1} = B_{k\Delta}$, $A_{k\Delta} = diag\left\{-c_{2,2}^T b_{kl,i}^N f_{kl,i}\right\} \in \mathbb{R}^{3N \times 6N}$, $A_{Rk} = diag\left\{A_{Rkl,i}(f_{kl,i})\right\} \in \mathbb{R}^{6N \times 6N}$, $B_{k\Delta 2} = diag\left\{c_{2,1}\left(b_{kl,i}^N\right)^{-1}\right\} \in \mathbb{R}^{6N \times 3N}$ $(l = 1,2,3)$, and $\Delta'_k(t) = \left[\Delta'_{ki}(t)\right] \in \mathbb{R}^{N \times 1}$. From (3.11) and (3.12), the error model can be rewritten as follows:

$$\dot{X}_k(t) = \hat{A}_k X_k(t) + \hat{B}_k \Delta'_k(t), k = p, \Theta, \tag{3.13}$$

where

$$\hat{A}_k = \begin{bmatrix} A_{kc} & B_{k\Delta} A_{k\Delta} \\ 0_{6N \times 6N} & A_{Rk} \end{bmatrix}, \hat{B}_k = \begin{bmatrix} B_{k\Delta 1} \\ B_{k\Delta 2} \end{bmatrix}.$$

Define $\lambda_{ekl}(t)$ and $\lambda_{ukl}(t)$ $(l = 0,1)$ as continuous and uniformly positive functions with upper bounds $\|\lambda_{ekl}\|_\infty$ and $\|\lambda_{ukl}\|_\infty$. The equivalent disturbances Δ'_p and Δ'_Θ are assumed to satisfy

$$\|\Delta'_k(t)\| \le \sum_{l=0}^{1} \lambda_{ekl} \|E(t - \rho_l(t))\| + \sum_{l=0}^{1} \lambda_{ukl} \|u_k(t - \rho_l(t))\| + \gamma_{kd},\ \ k = p, \Theta,$$

where $\lambda_{ekl} = \|\lambda_{ekl}\|_\infty$, $\lambda_{ukl} = \|\lambda_{ukl}\|_\infty$, and γ_{kd} is uniformly bounded positive function involving the external disturbance d_k, and

$$E(t) = \left[e_p^T(t),\ e_v^T(t),\ e_\Theta^T(t),\ e_\omega^T(t)\right]^T.$$

From (3.5) and (3.9), the equivalent disturbances Δ'_p and Δ'_Θ can be rewritten by

$$\begin{aligned} \|\Delta'_k(t)\| \le & \sum_{l=0}^{1} \lambda_{ekl} \|E(t-\rho_l(t))\| \\ & + \sum_{l=0}^{1} \lambda_{ukl}(\lambda_{kE} \|E(t-\rho_l(t))\| + f_{km} \|X_{Rk}(t-\rho_l(t))\|) + \gamma_{kd}, \end{aligned} \tag{3.14}$$

where $X_{Rk}(t) = diag(X_{Rk1}(t),\ X_{Rk2}(t),\ X_{Rk3}(t))$, $f_{km} = \|f_{ki}\|$, and $\lambda_{kE} = \|diag(K_p, K_v, K_\omega, K_\Theta)\|$. The communication delays satisfy $\bar{\rho}_e = \max_l \|\rho_l\|_\infty < \infty$ and $\bar{\rho}_{de} = \max_l \|\dot{\rho}_l\|_\infty < 1$. Let $\rho_0(t) = 0$. Denote P_k $(k = p, \Theta)$ as the solution to the Lyapunov equation: $P_k \hat{A}_k + \hat{A}_k^T P_k = -I_{12N}$. From (3.13), one can obtain that the matrix $\hat{A}_k$ is Hurwitz, and thereby P_k is positive definite. There exists a positive constant λ_{Bk} satisfying

$$\left\|P_k \hat{B}_k\right\| \le \lambda_{Bk} f_{km}^{-1}, k = p, \Theta.$$

Define

$$\xi_{ek} = \sum\nolimits_{l=0}^{1} \lambda_{Bk} \lambda_{ukl},$$

$$\xi_{efk} = \sum\nolimits_{l=0}^{1} \lambda_{Bk} \lambda_{ekl} + \sum\nolimits_{l=0}^{1} \lambda_{Bk} \lambda_{ukl} \lambda_{kE},$$

$$\xi_{uk} = \sum\nolimits_{l=0}^{1} \lambda_{Bk} \lambda_{ukl},$$

$$\xi_{ufk} = \sum\nolimits_{l=0}^{1} \lambda_{Bk} \lambda_{ekl} + \sum\nolimits_{l=0}^{1} \lambda_{Bk} \lambda_{ukl} \lambda_{kE},$$

$$\xi_{esfk} = 2\sum\nolimits_{l=0}^{1} \lambda_{Bk} \lambda_{ekl},$$

$$\xi_{usk} = 2\sum\nolimits_{l=0}^{1} \lambda_{Bk} \lambda_{ekl},$$

$$\xi_{\gamma k} = 2\lambda_{Bk} \gamma_{kd}.$$

Theorem 3.1

Consider the quadrotor dynamics given by (3.1), and the formation control protocol developed in Section 3.3. If $\hat{G}$ has a spanning tree, the information can flow from the virtual leader to the root, for a given bounded and piece-wise continuous initial state $E(\tau)$, $\tau \in [t_0 - \bar{\rho}_e,\ t_0]$, and a given initial time t_0, there exist constants $\underline{f_{ki}}(k = p,\Theta)$ $(k = p,\Theta)$ such that for any $f_{ki} > \underline{f_{ki}}$, $E_k(t)$ is uniformly bounded for $t \geq t_0$. Furthermore, for a given constant ε, there exists a constant $\underline{T}$ such that the state $E(t)$ satisfies $\|E(t)\| \leq \varepsilon,\ \forall t \geq \underline{T}$.

Proof 3.1 Consider the Lyapunov function candidate by neglecting the communication delays as

$$V_1(X(t)) = \sum_{k=p,\Theta} X_k^T(t) P_k X_k(t). \tag{3.15}$$

By differentiating (3.15), one has that

$$\begin{aligned}\dot{V}_1(X(t)) &= -\sum_{k=p,\Theta}\left(\|E_k(t)\|^2 + \|X_{Rk}(t)\|^2\right) + \sum_{k=p,\Theta} 2X_k^T(t)P_k\hat{B}_k\Delta_k(t)\\ &\leq -\sum_{k=p,\Theta}\left(\|E_k(t)\|^2 + \|X_{Rk}(t)\|^2\right)\\ &\quad + \sum_{k=p,\Theta} 2\left(\|E_k(t)\| + \|X_{Rk}(t)\|\right)\lambda_{Bk} f_{km}^{-1}\|\Delta_k'(t)\|.\end{aligned} \tag{3.16}$$

Substituting the inequality in (3.14) into (3.16), one can obtain that

$$\begin{aligned}\dot{V}_1(X(t)) \leq &-\sum_{k=p,\Theta}(1 - \xi_{ek} - \xi_{efk} f_{km}^{-1})\|E_k(t)\|^2\\ &-\sum_{k=p,\Theta}(1 - \xi_{uk} - \xi_{ufk} f_{km}^{-1})\|X_{Rk}(t)\|^2\\ &+\sum_{k=p,\Theta}\xi_{esfk}\|E_k(t - \rho_l(t))\|^2 f_{km}^{-1}\\ &+\sum_{k=p,\Theta}\xi_{\gamma k}\|E_k(t)\| f_{km}^{-1} + \sum_{k=p,\Theta}\xi_{\gamma k}\|X_{Rk}(t)\| f_{km}^{-1}\\ &+\sum_{k=p,\Theta}\xi_{usk}\|X_{Rk}(t - \rho_l(t))\|^2.\end{aligned} \tag{3.17}$$

Then, by introducing the communication delays, one can obtain the final Lyapunov function candidate as

$$V(X(t),t)= V_1(X(t)) + \sum_{k=p,\Theta} \sum_{l=0}^{1} \xi_{NR} \int_{t-\rho_l(t)}^{t} \|X_{Rk}(\tau)\|^2 d\tau$$
$$+ \sum_{k=p,\Theta} \sum_{l=0}^{1} \xi_{NE} \int_{t-\rho_l(t)}^{t} \|E_k(\tau)\|^2 d\tau,$$

where ξ_{NE} and ξ_{NR} are positive constants and satisfy $1-2\xi_{NE} > \xi_{ek}, 1-2\xi_{NR} > \xi_{uk}$, and $2(1-\bar{\rho}_{de})\xi_{NR} > \xi_{usk}$. By differentiating the final Lyapunov function candidate, one can obtain that

$$\dot{V}(X(t),t) \le \dot{V}_1(X(t))$$
$$+ \sum_{k=p,\Theta} \sum_{l=0}^{1} \xi_{NE} \left(\|E_k(t)\|^2 - (1-\bar{\rho}_{de}) \|E_k(t-\rho_l(t))\|^2 \right) \tag{3.18}$$
$$+ \sum_{k=p,\Theta} \sum_{l=0}^{1} \xi_{NR} \left(\|X_{Rk}(t)\|^2 - (1-\bar{\rho}_{de}) \|X_{Rk}(t-\rho_l(t))\|^2 \right).$$

Let $\pi_{ek} = 1-2\xi_{NE} - \xi_{ek} - \xi_{efk} f_{km}^{-1}$, $\pi_{Rk} = 1-2\xi_{NR} - \xi_{uk} - \xi_{ufk} f_{km}^{-1}$, $\pi_{esk} = 2(1-\bar{\rho}_{de})\xi_{NE} - \xi_{esfk} f_{km}^{-1}$, $\pi_{Rsk} = 2(1-\bar{\rho}_{de})\xi_{NR} - \xi_{usk}$, and $\pi_{\gamma k} = \xi_{\gamma k} f_{km}^{-1}$. Substituting (3.17) into (3.18), one can obtain that

$$\dot{V}(X(t),t) \le -\sum_{k=p,\Theta} \pi_{ek} \|E_k(t)\|^2 - \sum_{k=p,\Theta} \pi_{Rk} \|X_{Rk}(t)\|^2$$
$$- \sum_{k=p,\Theta} \pi_{Rsk} \|X_{Rk}(t-\rho_d(t))\|^2 - \sum_{k=p,\Theta} \pi_{esk} \|E_k(t-\rho_d(t))\|^2 \tag{3.19}$$
$$+ \sum_{k=p,\Theta} \pi_{\gamma k} \|E_k(t)\| + \sum_{k=p,\Theta} \pi_{\gamma k} \|X_{Rk}(t)\|.$$

One can see that if f_{km} $(k = p,\Theta)$ satisfy that

$$f_{km} > \xi_{efk} / (1-2\xi_{NE} - \xi_{ek}),$$
$$f_{km} > \xi_{ufk} / (1-2\xi_{NR} - \xi_{uk}),$$
$$f_{km} > \xi_{esfk} / 2(1-\bar{\rho}_d)\xi_{NE}, k = p,\Theta,$$

then π_{ek} and π_{Rk} are positive and π_{Rsk}, π_{esk}, and $\pi_{\gamma k}$ are nonnegative. From (3.19), one can obtain that

$$\begin{aligned}\dot{V}(X(t),t) \leq & -\sum_{k=p,\Theta} \pi_{ek}\left\|E_k(t)\right\|^2 + \sum_{k=p,\Theta} \pi_{\gamma k}\left\|E_k(t)\right\| \\ & -\sum_{k=p,\Theta} \pi_{Rk}\left\|X_{Rk}(t)\right\|^2 + \sum_{k=p,\Theta} \pi_{\gamma k}\left\|X_{Rk}(t)\right\| \\ \leq & -\sum_{k=p,\Theta} (\pi_{ek} - \pi_{\gamma k})\left\|E_k(t)\right\|^2 \\ & -\sum_{k=p,\Theta} (\pi_{Rk} - \pi_{\gamma k})\left\|X_{Rk}(t)\right\|^2 + \pi_{\gamma k}/2.\end{aligned} \tag{3.20}$$

From (3.20), one can obtain that the attractive radius of $E(t)$ is determined by $\pi_{\gamma k}/2$. In fact, if f_{km} are sufficiently large, $\pi_{\gamma k}$ can be made as small as possible and $\pi_{ek} - \pi_{\gamma k} > 0$ and $\pi_{Rk} - \pi_{\gamma k} > 0$. Therefore, one can obtain that Theorem 3.1 holds. □

Remark 3.5

It should be pointed out that the theoretical values of the robust filter parameters f_{ki} $(k = p, \Theta)$ determined by Theorem 3.1 may be conservative, that is, the actual values of f_{ki} $(k = p, \Theta)$ may be much smaller than their theoretical values. Since the tracking performances can be improved by selecting f_{ki} $(k = p, \Theta)$ with larger values, the robust filter parameters can be tuned online unidirectionally following this procedure. The first step is to set $f_{\Theta i}$ with a certain initial positive value. The second step is to increase the value of $f_{\Theta i}$ until the achieved attitude tracking performances are satisfied. The third step is to determine f_{pi} in a similar way.

3.5 Experimental Results

In this section, the experimental results of a team of micro quadrotors are provided to validate the effectiveness of the proposed robust formation control approach. The main onboard sensors installed at each quadrotor consist of a decentralized localization system using ultra-wideband radio triangulation (UWB deck), an inertial measurement unit module including a 3-D gyroscope, a 3-D accelerometer with 3-D magnetometer, a laser sensor measuring the altitude of the quadrotor, and a Flow Deck measuring movements in relation to the ground. The data from the inertial measurement unit and magnetometer sensor are fused to estimate the attitude angles and angle rates of the quadrotors. The positioning system used in the experiment is a local positioning system named UWB positioning system, and it is used to provide the current position of each quadrotor in the formation flights as depicted in [30]. The UWB system consists of a set of anchors and tags. The anchors are used as the position reference, and the tags measure the distance from each anchor to the tags. The information needed to calculate the position is available in the tag which enables the position estimation on board of each micro quadrotor. The UWB deck contains tags, which are attached to the micro quadrotors expansion board and serve as the reference base for the positioning system. Typically, 2.4 GHz wireless modules are used for the communication between the formation micro quadrotors and the ground client. The configuration of the experimental system is depicted in Figure 3.1.

The nominal parameters of the trajectory tracking and attitude controller gains K_p, K_v, K_q, and K_ω are tuned step by step as shown in the last paragraph in Section 3.4. The nominal controller parameters are selected by the following steps. First, μ_p, K_p, and K_v are determined by using the LQR-based control method. Second, K_Θ and K_ω are chosen based on the LQR approach for better attitude performance of each quadrotor. The nominal parameters are selected as $\mu_p = 3$, $K_p = diag\{1,1,1\}$, $K_v = \{0.5,0.5,0.5\}$, $K_\Theta = \{30,30,30\}$, and $K_\omega = diag\{15,15,15\}$. The robust compensator parameters are selected as $f_{p,i} = diag\{1,1,1\}$ and $f_{\Theta,i} = diag\{15,15,15\}$ $(i = 1,2,3)$.

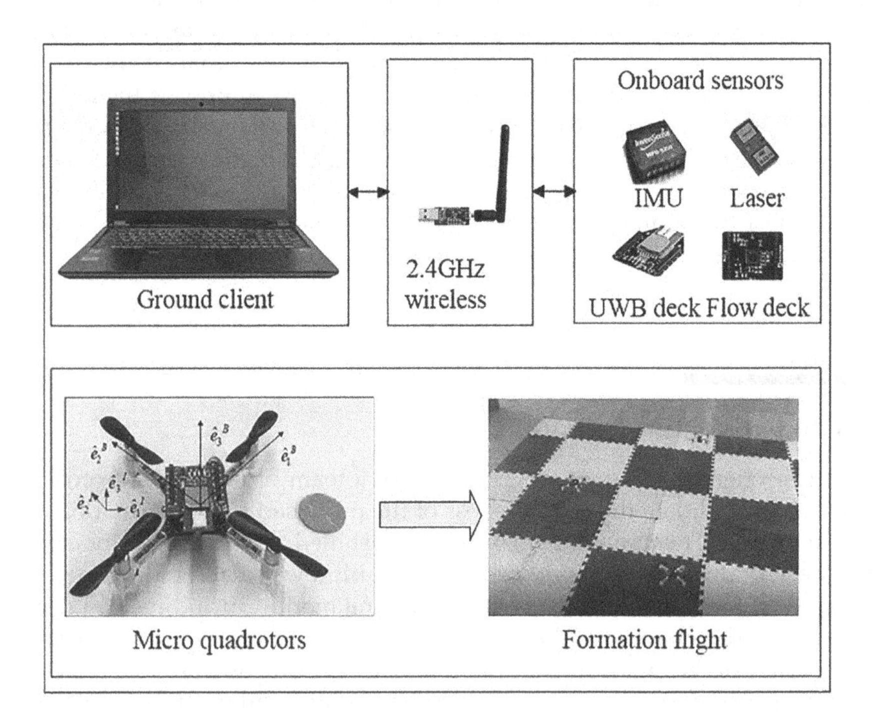

FIGURE 3.1
Formation hardware system.

Based on the formation control protocol designed in the third section, three micro quadrotors are used in the time-varying formation flight experiments. The positions of the three micro quadrotors are required to realize geometrical configurations, and the formation center tracks the reference trajectory. The initial geometrical configuration is selected as $\varsigma_{10} = \begin{bmatrix} -0.6 & 0.6 & 0 \end{bmatrix}^T$, $\varsigma_{20} = \begin{bmatrix} -0.9 & 0.9 & 0 \end{bmatrix}^T$, and $\varsigma_{30} = \begin{bmatrix} -0.9 & 0.3 & 0 \end{bmatrix}^T$ m, and the ending formation geometrical configuration as $\varsigma_{1t} = \begin{bmatrix} -0.6 & 0.6 & 0 \end{bmatrix}^T$, $\varsigma_{2t} = \begin{bmatrix} 0.9 & 0.9 & 0 \end{bmatrix}^T$, and $\varsigma_{3t} = \begin{bmatrix} 0.9 & 0.3 & 0 \end{bmatrix}^T$ m. The reference trajectory of the formation center is a rectangle with $O_{p1} = \begin{bmatrix} -0.6 & 0.6 & 0.6 \end{bmatrix}$, $O_{p2} = \begin{bmatrix} 0.6 & 0.6 & 0.6 \end{bmatrix}$, $O_{p3} = \begin{bmatrix} 0.6 & -0.6 & 0.6 \end{bmatrix}$, and $O_{p4} = \begin{bmatrix} -0.6 & -0.6 & 0.6 \end{bmatrix}$ m as the four vertexes. The yaw angle reference ψ_i^r of each quadrotor in the formation is set to be 0. The directed graph $G = (V, E, W)$ is modeled to describe the information among three vehicles, where $V = \{v_1, v_2, v_3\}$, $E = \{(v_1, v_2), (v_2, v_3)\}$, and $W = \begin{bmatrix} w_{ij} \end{bmatrix}$ with $w_{ij} = 1$ for $(v_i, v_j) \in E$ and 0 otherwise. Only quadrotor 1 can obtain the information from the virtual team leader, and thereby $\alpha_{p1} = 1$, $\alpha_{p2} = 0$, and $\alpha_{p3} = 0$. The sampling rate of a communication sensor is 20 Hz. Therefore, one can obtain that the communication delays satisfy $\|\rho_{kj}(t)\|_\infty < 0.05$ s $(k = p, \Theta)$ and $\bar{\rho}_e \le 0.05$ s.

In these experimental results, dark gray, light gray, and black solid lines represent quadrotors 1–3, and the dotted line indicates the formation pattern. Figure 3.2 depicts the 3-D trajectory and formation geometrical configurations of the three quadrotors. In Figures 3.3 and 3.4, the longitudinal, lateral,

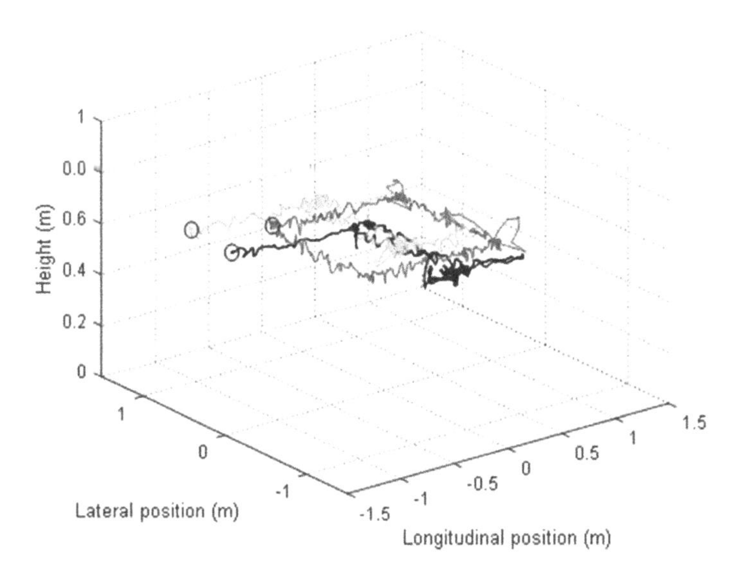

FIGURE 3.2
Three-dimensional trajectories by the proposed controller.

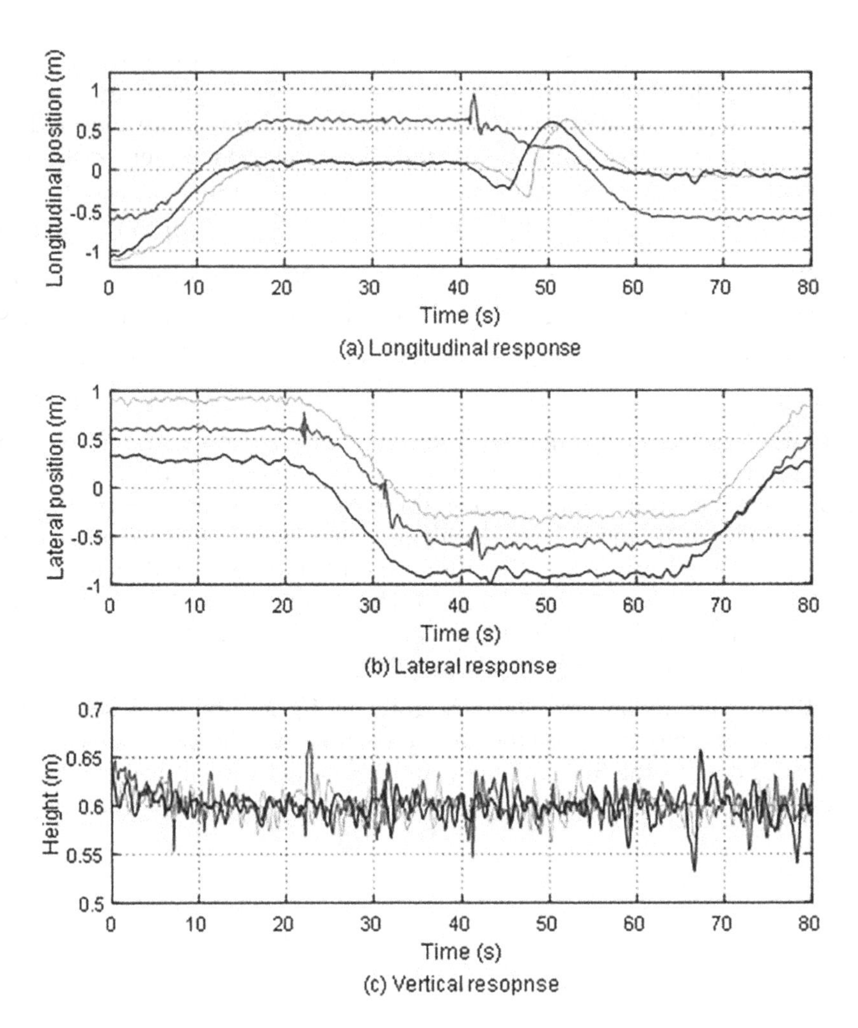

FIGURE 3.3
Position response by the proposed controller.

and height position response and tracking errors are given, respectively. Figure 3.5 depicts the Euler angle response for the intuitive description. Then, a baseline controller developed from [32] is introduced for comparisons, and its position tracking errors are given in Figure 3.6. It can be seen that using the proposed method, the longitudinal, lateral, and height errors are nearly 0.09, 0.08, and 0.04 m, respectively. However, by the introduced

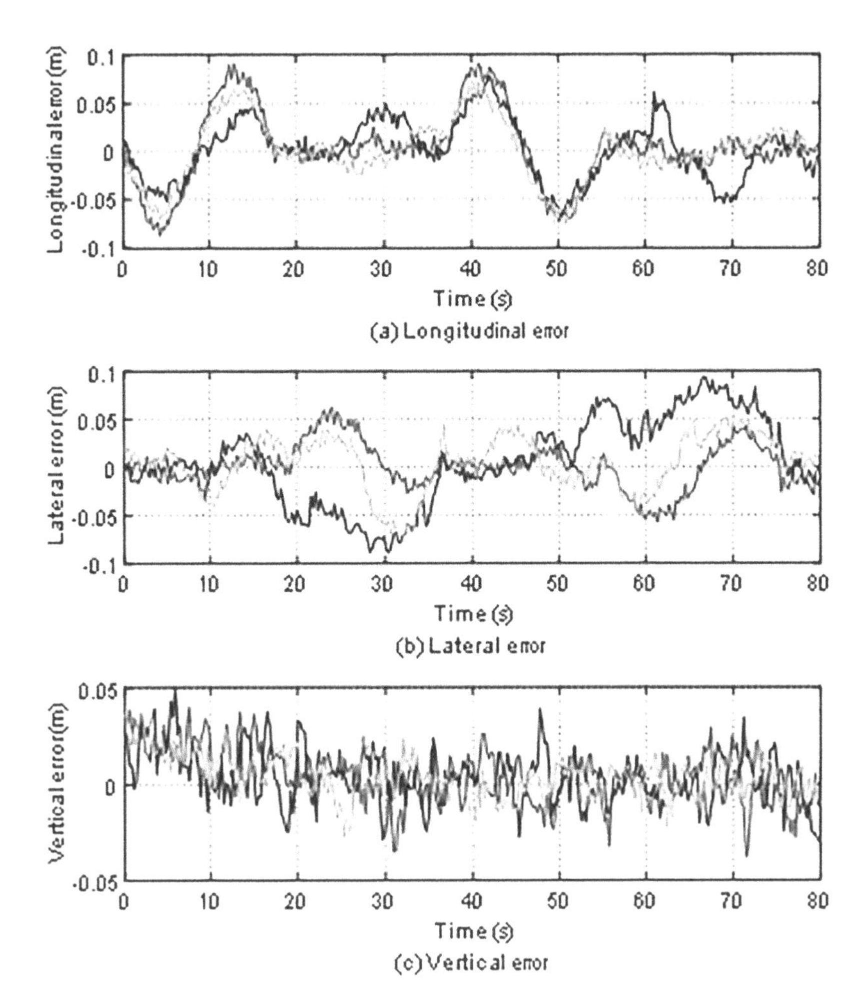

FIGURE 3.4
Trajectory tracking errors by the proposed controller.

leader-following controller, the longitudinal, lateral, and height errors are about 0.20, 0.25, and 0.08 m, respectively. It can be observed that the proposed global closed-loop control system can *improve* the formation tracking performance compared to the baseline controller. Besides, the effects of nonlinear dynamics, parametric perturbations, external disturbances, and communication delays can be restrained by the proposed formation protocol.

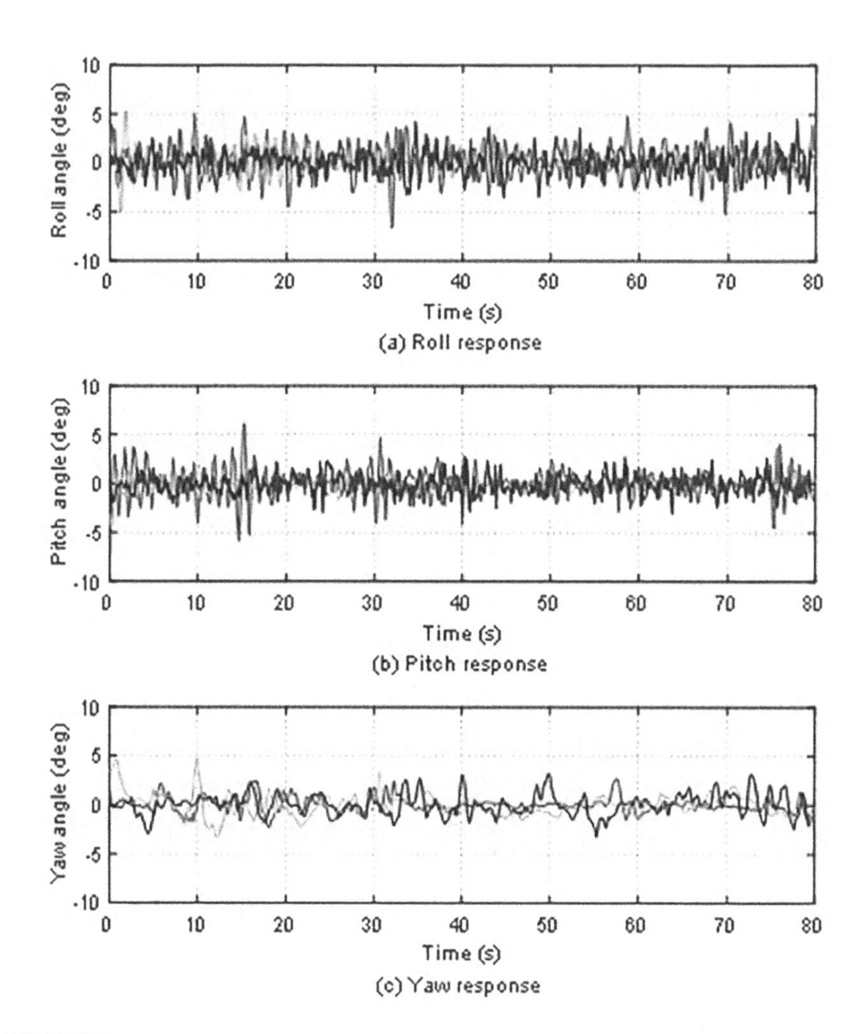

FIGURE 3.5
Attitude responses by the proposed controller.

3.6 Conclusion

This chapter presents a robust formation trajectory tracking controller design method to address the formation trajectory tracking control problem for a team of uncertain quadrotors subject to communication delays. For each micro quadrotor, a robust controller is developed consisting of a trajectory tracking controller and an attitude controller. Robustness properties are

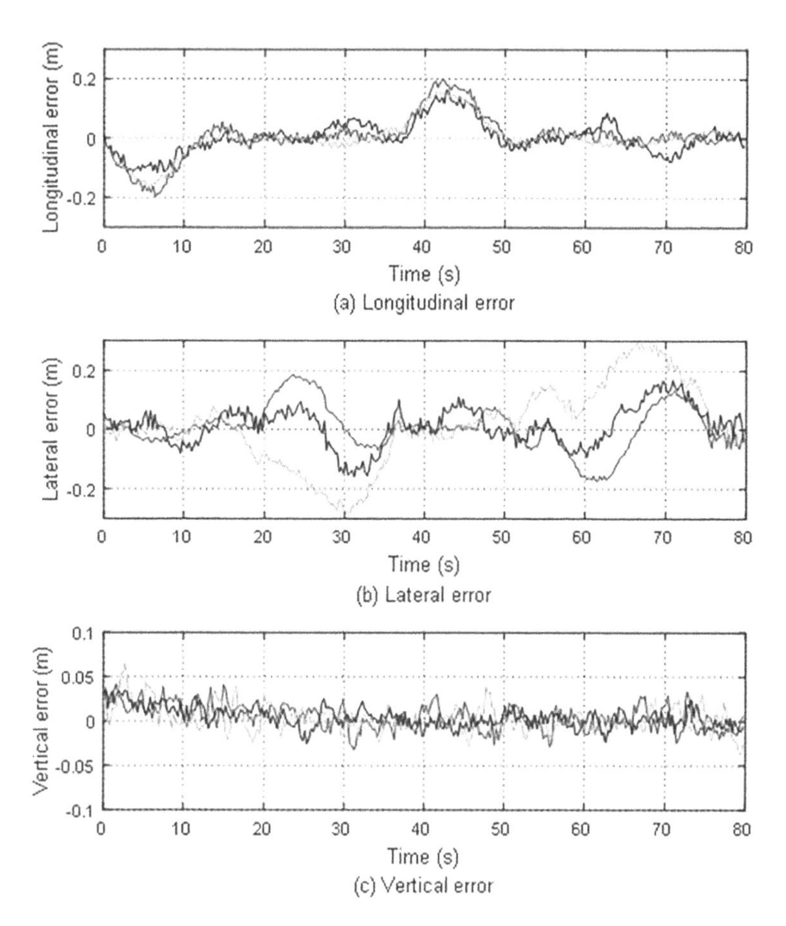

FIGURE 3.6
Trajectory tracking errors by a baseline controller.

analyzed, and the tracking errors of the global closed-loop control system are proven to converge into a given neighborhood of the origin in a finite time. Experimental results demonstrate the advantages of the proposed formation control protocol.

4

Robust Formation Tracking Control for Multiple Quadrotors Subject to Switching Topologies

In this chapter, the formation control problem for a team of quadrotors subject to switching topologies is studied. The quadrotor model is underactuated and includes nonlinear dynamics, parameter uncertainties, and external disturbance. A distributed robust controller is developed, which consists of a position controller to govern the translational motion for the desired formation and an attitude controller to control the rotational motion of each quadrotor. Theoretical foundations and detailed simulation tests demonstrate the effectiveness of the proposed method.

4.1 Introduction

Research activities in the formation flight of unmanned aerial vehicles (UAVs) have increased substantially in the last few years. UAV formation flight can potentially be useful for increased surveillance coverage, better target acquisition, and increased security measures, as shown in [63–67].

Several studies investigate fixed communication topologies of multi-agent/multi-UAV systems, modeled by first- and/or second-order differential equations. In [68], a graph theory method with necessary and sufficient conditions to reach consensus with reference velocity was developed. In [24], an adaptive output-feedback method was proposed to address the distributed time-varying formation control problem. In [69], the time-invariant formation control problem for multi-agent systems was investigated. A robust formation protocol based on leader-follower was developed by combining a relative localization strategy and a complex Laplacian-based formation control method [70]. A distributed control method designed by utilizing an artificial potential function was presented to guarantee formation performance [71]. In [72], a finite-time consensus problem for second-order nonlinear multi-agent systems under communication constraints was investigated. However, the dynamical models in above mentioned were linear; the formation control problems of nonlinear systems were not further discussed.

DOI: 10.1201/9781003242147-4

In [73], a leader-follower decentralized controller approach for multiple mobile robots was illustrated to achieve formation tracking. In [74], a distributed adaptive control method based on backstepping was developed. In [14], the formation control problem of nonlinear multi-agent systems was discussed, and a distributed formation control method was proposed. However, in [73–75], parameter uncertainties and external disturbance were not fully analyzed in the closed-loop control systems. In [16], a robust formation control protocol was presented and a suboptimal controller was designed to restrain model parameter uncertainties and external disturbances. But this method cannot suppress the influence of uncertainties in the whole frequency range as desired. In [68–75], formation control problems were discussed mainly based on fixed communication topologies, without considering switching topologies.

It should be pointed out that in the above text, only formation stabilization problems for multi-UAV systems were addressed. In many practical applications, such as source seeking and target enclosing, forming the desired time-varying formation is only the first step for a multi-UAV system. In such scenarios, time-varying formation tracking problems arise, where a group of followers keeps the desired time-varying formation while tracking the trajectory of the leader. Considering the fact that the interaction topology among the agents may be unreliable, it is more meaningful to study time-varying formation tracking problems for multi-agent systems with switching interaction topologies.

However, formation control problems with fixed topologies are relatively easy to address, while problems with switching topologies are more challenging. In [76], a simple controllability condition for the network subject to switching topologies was derived to indicate that the controllability of the entire network did not depend on each specific topology of the network. A consistent strategy for time-varying reference signals in multi-vehicle systems was studied in [68]. The necessary and sufficient conditions for a linear multi-agent system to realize time-varying formation control are given in [77]. In [11], the time-varying formation control tracking problem was studied subject to switched topologies for second-order nonlinear multi-agent systems.

In this chapter, a robust formation control protocol is proposed for multi-UAV teams subject to switching topologies. The control law consists of a position and an attitude controller, which are both composed of a nominal controller designed to achieve desired tracking for the nominal system, and a disturbance estimating controller designed to restrain the effects of uncertainties and nonlinear dynamics.

Compared with the previous relevant results, the contributions of this chapter are threefold. First, the considered quadrotor model is underactuated and has complete nonlinear and coupled dynamics. This model is more reliable and accurate when tackling quadrotor attitude and position changes from the perspective of dynamics. However, the quadrotor models were reduced to second-order linear models, which cannot accurately describe quadrotor dynamics in [68–72]. Second, the interaction topology among the quadrotors can be switched. The constructed multiple quadrotor systems that can achieve time-varying formation tracking are both necessary and sufficient. However, in [68, 70, 71, 72, 74, 75], formation control issues were studied based on fixed communication topologies. Third, the effects of parametric uncertainties, nonlinear and coupled dynamics, and external disturbances can be restrained simultaneously, which could affect the robustness properties and tracking performance of the overall control system. In [72, 74–76], uncertainty rejection problems were not fully discussed.

4.2 Preliminaries and Problem Description

4.2.1 Graph Theory

Let $G = \{Q, E, W\}$ denote a weighted directed graph that describes the information exchange among n quadrotors. $Q = \{q_1, q_2, \ldots, q_n\}$ indicates a set of n nodes, where q_i represents the micro quadrotor i. The influence relationship of the quadrotor i on the quadrotor j corresponds to one edge $e_{ij} = (q_i, q_j)$ of G, which means that quadrotor i can access the information from quadrotor j. Let $\Phi = \{1, 2, \ldots, n\}$. All the influence relationships between the quadrotors correspond to an edge set $E \subseteq Q \times Q$, and $W = [w_{ij}] \in \mathbb{R}^{n \times n}$ is the associated adjacency matrix describing the strength of influence between the quadrotors. w_{ij} represents the weight value between the quadrotor i and the quadrotor j, which is the strength of influence between quad-copter i and quad-copter j. For any $i, j \in \{1, 2, \ldots, N\}$, $w_{ij} > 0$ only if $(q_i, q_j) \in E$, and $w_{ij} = 0$ otherwise. The neighbor set of node i is depicted by $N_i = \left\{q_j \in V : (q_j, q_i) \in E\right\}$. Define the in-degree matrix as $D = diag\{d_i\} \in \mathbb{R}^{N \times N}$ with $\deg_{in}\{v_i\} = \sum_{j=1}^{N} w_{ij}$. The graph Laplacian matrix is defined as $L = D - W$. A sequence of ordered edges with the form of $\left\{(q_i, q_a), (q_a, q_b), \ldots, (q_k, q_j)\right\}$ gives a direct path from quadrotor i to quadrotor j. If there exists a subset of the edges having a path from a node, called the root, to all the other nodes, G is said to have a spanning tree.

The directed graph G discsued in this chapter can be switching, and the graph Laplacian matrix as L is time varying. Define the total formation flight time as t^F. The finite time interval $[0, t^F)$ can be divided as a set of bounded intervals $[t_m, t_{m+1})(m \in \Xi)$, where Υ represents all natural numbers. The dwell time τ satisfies $0 < \tau \leq (t_{m+1} - t_m) \leq t^F$, and the directed graph is fixed for the bounded interval $[t_m, t_{m+1})$. Let $\sigma(t)$ represent a switching signal. Define $W_{\sigma(t)} = [w_{\sigma(t)}^{ij}] \in \mathbb{R}^{n \times n}$ and $N_i^{\sigma(t)} = \left\{j \middle| (q_i, q_j) \in E\right\}$ as the weighted adjacency matrix and the set of neighbors of quadrotor i at $\sigma(t)$. Let $G_{\sigma(t)}$ and $L_{\sigma(t)} = D_{\sigma(t)} - W_{\sigma(t)}$

denote the directed graph at t and the Laplacian matrix at $\sigma(t)$, respectively, where $D_{\sigma(t)} = diag\{\sum_{j=1}^{n} w_{\sigma(t)}^{ij}\}$. If there is at least one node that have a path to all other nodes, the directed graph $G_{\sigma(t)}$ has a spanning tree.

4.2.2 System Model

Consider N quad-copters modeled as rigid bodies. As depicted in [51], let $\mathbb{Z}_I = \{\mathbb{Z}_{Ix}, \mathbb{Z}_{Iy}, \mathbb{Z}_{Iz}\}$ denote the earth-fixed inertial frame and $\mathbb{Z}_{Bi} = \{\mathbb{Z}_{Bxi}, \mathbb{Z}_{Byi}, \mathbb{Z}_{Bzi}\}$ the body-fixed frame attached to quadrotor i with the origin in its mass center. Let $\eta_{Ii} = [p_{Ii}^T \quad v_{Ii}^T]^T \in \mathbb{R}^{6\times1}$ represent the position and linear velocity of quadrotor i in $\mathbb{Z}_I$, where $p_{Ii} = [p_{Ixi} \quad p_{Iyi} \quad p_{Izi}]^T$ and $v_{Ii} = [v_{Ixi} \quad v_{Iyi} \quad v_{Izi}]^T$. The attitude of quadrotor i is described by the four-element unit quaternion $q_i = [q_{0i} \quad \bar{q}_i]^T \in \mathbb{R}^{4\times1}$ to avoid the singularity, which satisfies $q_{0i}^2 + \|\bar{q}_i\|^2 = 1$ and $\bar{q}_i = [q_{1i} \quad q_{2i} \quad q_{3i}]^T \in \mathbb{R}^{3\times1}$. The coordinate transformation matrix $R(q_i) \in SO(3)$, which brings $\mathbb{Z}_{Bi}$ to $\mathbb{Z}_I$ and satisfies $R(q_i)R^T(q_i) = I_3$, can be given by

$$R(q_i) = \begin{bmatrix} 1-2q_{2i}^2 & 2q_{1i}q_{2i} - 2q_{0i}q_{3i} & 2q_{1i}q_{3i} + 2q_{0i}q_{2i} \\ 2q_{1i}q_{2i} + 2q_{0i}q_{3i} & 1-2q_{1i}^2 - 2q_{3i}^2 & 2q_{2i}q_{3i} - 2q_{0i}q_{1i} \\ 2q_{1i}q_{3i} - 2q_{0i}q_{2i} & 2q_{2i}q_{3i} + 2q_{0i}q_{1i} & 1-2q_{1i}^2 - 2q_{2i}^2 \end{bmatrix}.$$

Each quadrotor model can be divided into a translational dynamics and a rotational dynamics. From [51], the translational dynamics of quadrotor can be given as

$$\begin{aligned} \dot{p}_{Ii} &= v_{Ii}, \\ \dot{v}_{Ii} &= \beta_{fi} R(q_i) u_{0i} c_{3,3} - g c_{3,3} + d_{T,fi}, \end{aligned} \tag{4.1}$$

and rotational dynamics as

$$\begin{aligned} \dot{R}(q_i) &= R(q_i) S(\omega_i), \\ \dot{\omega}_i &= -J_i^{-1} S(\omega_i) J_i \omega_i + \beta_{\tau i} \bar{u}_i + d_{R,\tau i}, \end{aligned} \tag{4.2}$$

where $\omega_i = [\omega_{xi} \quad \omega_{yi} \quad \omega_{zi}]^T$ is the angular velocity of quadrotor i in $\mathbb{Z}_{Bi}$, $\bar{u}_i = \begin{bmatrix} u_{1i} & u_{2i} & u_{3i} \end{bmatrix}^T$, g is the acceleration of gravity, $J_i = diag\{J_{1i}, J_{2i}, J_{3i}\}$ is the inertia matrix in $\mathbb{Z}_{Bi}$, $d_{T,fi}$ and $d_{R,\tau i}$ represent the influence of external disturbance on the translational dynamics and rotational dynamics of quadrotor i in $\mathbb{Z}_I$ and $\mathbb{Z}_{Bi}$, respectively, and $\beta_{fi} = k_{fi} m_i^{-1} I_3$ and $\beta_{\tau i} = J_i^{-1} diag\{l_{mc} k_{fi}, l_{mc} k_{fi}, k_{\tau i}\}$, with k_{fi}, $k_{\tau i}$, l_{mc}, and m_i denote the model parameters. For quadrotor i, the inputs $u_i = [u_{0i} \quad u_{1i} \quad u_{2i} \quad u_{3i}]^T \in \mathbb{R}^{4\times1}$ can be described as follows:

$$u_{0i} = \omega_{R,1i}^2 + \omega_{R,2i}^2 + \omega_{R,3i}^2 + \omega_{R,4i}^2,$$
$$u_{1i} = \omega_{R,2i}^2 - \omega_{R,4i}^2,$$
$$u_{2i} = \omega_{R,1i}^2 - \omega_{R,3i}^2,$$
$$u_{3i} = \omega_{R,1i}^2 - \omega_{R,2i}^2 + \omega_{R,3i}^2 - \omega_{R,4i}^2,$$

where $\omega_{R,ji}(j = 1,2,3,4)$ represents the rotation velocities of rotors.

Remark 4.1

It can be observed that the quadrotor dynamical system is underactuated and the quadrotor dynamics involves highly nonlinearity and coupling.

In practical application, it is difficult to obtain the accurate values of the vehicle parameters. Let the subscript N be the nominal parameter and the subscript Δ the parameter perturbations satisfying $J_i = J_i^N + J_i^\Delta$, $\beta_{fi} = \beta_{fi}^N + \beta_{fi}^\Delta$, and $\beta_{\tau i} = \beta_{\tau i}^N + \beta_{\tau i}^\Delta$. Then, the nonlinear system model (4.1) and (4.2) can be rewritten as follows:

$$\begin{aligned} \dot{p}_{Ii} &= v_{Ii}, \\ \dot{v}_{Ii} &= \beta_{fi}^N F_i - g c_{3,3} + \Delta_{Ti}, \end{aligned} \tag{4.3}$$

and

$$\begin{aligned} \dot{R}(q_i) &= R(q_i) S(\omega_i), \\ \dot{\omega}_i &= -\left(J_i^N\right)^{-1} S(\omega_i) J_i^N \omega_i + \beta_{\tau i}^N \bar{u}_i + \Delta_{Ri}, \end{aligned} \tag{4.4}$$

where $q_i^r = [q_{0i}^r \quad q_{1i}^r \quad q_{2i}^r \quad q_{3i}^r]^T$ represents desired reference of quaternions, $F_i = R(q_i^r) u_{0i} c_{3,3}$ is the virtual position control input, $\tilde{F}_i = \left[\tilde{F}_{1i} \quad \tilde{F}_{2i} \quad \tilde{F}_{3i}\right]^T = u_{0i} (\beta_{fi}^N)^{-1} \beta_{fi} R(q_i) c_{3,3} - F_i$, and Δ_{Ti} and Δ_{Ri} are called the equivalent disturbance satisfying

$$\Delta_{Ti} = \beta_{fi}^N \tilde{F}_i + d_{T,fi},$$
$$\Delta_{Ri} = (J_i^N)^{-1} S(\omega_i) J_i^N \omega_i - J_i^{-1} S(\omega_i) J_i \omega_i + \beta_{\tau i}^\Delta \bar{u} + d_{R,\tau i}.$$

The nonlinear model (4.3) and (4.4) indicates the real model of each quadrotor, and thus the nominal model can be obtained by ignoring Δ_{Ti} and Δ_{Ri} in (4.3) and (4.4). In fact, the quadrotor formation model is obtained to design the proposed robust formation controller.

4.2.3 Problem Description

The objective of this chapter is to design a distributed formation control protocol for each quadrotor in the team such that the formation center of the quadrotors tracks a prescribed stationary trajectory and the group of quadrotors keep time-varying geometrical configurations. Define $p_0^r \in \mathbb{R}^{3\times1}$ as the reference trajectory of the formation center. p_0^r is differentiable, satisfying $\dot{p}_0^r = v_0^r$ and $\ddot{p}_0^r = 0_{3\times1}$. Let $\delta_{ij} = [\delta_{xij} \quad \delta_{yij} \quad \delta_{zij}]^T$ be the desired position deviation between the quadrotor i and quadrotor j. In addition, the desired position deviation δ_{ij} satisfies $\delta_{ij} = \delta_i - \delta_j$ $(i, j \in \Phi)$, where δ_i indicates the position deviation between the virtual leader.

4.3 Formation Control Protocol Design

For each quadrotor, the proposed robust controller subject to switching topologies consists of a trajectory tracking controller to track the desired formation trajectory and form the time-varying geometrical configurations, and an attitude controller to govern the attitude angles in the rotational motion. Both the position and attitude controllers are divided into two parts: the nominal part to achieve the desired tracking of the nominal system and the disturbance estimating part to restrain the influence of uncertainties on the real system.

4.3.1 Position Controller Design

The virtual position control input F_i is designed with two parts as

$$F_i = F_i^N + F_i^\Delta, \tag{4.5}$$

where $F_i^N \in \mathbb{R}^{3\times1}$ denotes the nominal position control input and $F_i^\Delta \in \mathbb{R}^{3\times1}$ denotes the disturbance estimating input. The nominal trajectory tracking controller F_i^N subject to switching topologies is constructed based on a feedback approach as follows:

$$\begin{aligned} F_i^N = & -\alpha_F \sum_{j\in N_i^{\sigma(t)}} w_t^{ij} (\beta_{fi}^N)^{-1} K_p (p_{Ii} - p_{Ij} - \delta_{ij}) \\ & -\alpha_F \sum_{j\in N_i^{\sigma(t)}} w_t^{ij} (\beta_{fi}^N)^{-1} K_v (v_{Ii} - \dot{p}_{Ij}) \\ & -\alpha_F \beta_{Ii}^t (\beta_{fi}^N)^{-1} K_p (p_{Ii} - \delta_i - p_0^r) \\ & -\alpha_F \beta_{Ii}^t (\beta_{fi}^N)^{-1} K_v (v_{Ii} - \dot{p}_0^r) + (\beta_{fi}^N)^{-1} g c_{3,3}, \end{aligned} \tag{4.6}$$

where α_F denotes a positive scalar coupling gain, $K_p, K_v \in \mathbb{R}^{3\times3}$ denotes diagonal nominal controller parameter matrices, and β_{li}^t represents the time-varying connection weight between the virtual leader and quadrotor i. $\beta_{li}^t > 0$ indicates that the virtual team leader can send information to quad-copter i, otherwise $\beta_{li}^t = 0$.

Remark 4.2

The nominal control input F_i^N is designed based on the feedback linearization technique to restrain the influences of the nonlinearities on the nominal closed-loop control system and to achieve the desired formation trajectory tracking and the desired formation pattern for the nominal translational system.

Furthermore, the disturbance estimator F_i^Δ can be designed as follows:

$$F_i^\Delta = -(\beta_{fi}^N)^{-1}\Gamma_{pi}(s)\Delta_{Ti}, \tag{4.7}$$

where $\Gamma_{pi}(s) = diag\{\Gamma_{pj,i}(s)\} = diag\{\eta_{pj,i}^2 / (s+\eta_{pj,i})^2\}$ denotes the robust filters, with $\eta_{pj,i}$ denotes a positive filter parameter. However, Δ_{Ti} in (4.7) cannot be measured directly. Therefore, from (4.3), one can have that

$$\Delta_{Ti} = \ddot{p}_{li} - \beta_{fi}^N F_i + g c_{3,3}. \tag{4.8}$$

Substituting (4.7) into (4.8) yields the following realization for F_i^Δ

$$\begin{aligned} \dot{\vartheta}_{1i}^p &= -\eta_{pi}\vartheta_{1i}^p - \eta_{pi}^2 p_{li} + \beta_{fi}^N F_i - g c_{3,3}, \\ \dot{\vartheta}_{2i}^p &= -\eta_{pi}\vartheta_{2i}^p + 2\eta_{pi} p_{li} + \vartheta_{1i}^p, \\ F_i^\Delta &= \left(\beta_{fi}^N\right)^{-1} \eta_{pi}^2(\vartheta_{2i}^p - p_{li}), \end{aligned} \tag{4.9}$$

where $\vartheta_{1i}^p, \vartheta_{2i}^p \in \mathbb{R}^{3\times1}$ and $\eta_{pi} = diag\{\eta_{pj,i}\} \in \mathbb{R}^{3\times3}$ are the robust filter states.

Remark 4.3

One can see that the robust filters $\Gamma_{pj,i}(s)$ $(j = 1,2,3;\ i \in \Phi)$ have the property: if the filter parameter $\eta_{pj,i}$ is larger, the frequency bandwidth of $\Gamma_{pj,i}(s)$ would be wider within which the filter gain would get closer to 1. In this case, one can observe that F_i^Δ could follow the equivalent disturbance better and more influences of Δ_{Ti} on the real system could be restrained.

4.3.2 Attitude Controller Design

The attitude error, describing the discrepancy between the real quad-copter attitude and the desired one, can be obtained by a nonlinear function $e_{qi} = \tilde{Q}(q_i, q_i^r)$ satisfying

$$\tilde{Q}(q_i, q_i^r) = 2\,\mathrm{sgn}\left(\sum_{j=0}^{3} q_{ji}^r q_{ji}\right)\begin{bmatrix} -q_{0i}^r q_{1i} + q_{1i}^r q_{0i} + q_{2i}^r q_{3i} - q_{3i}^r q_{2i} \\ -q_{0i}^r q_{2i} - q_{1i}^r q_{3i} + q_{2i}^r q_{0i} + q_{3i}^r q_{1i} \\ -q_{0i}^r q_{3i} + q_{1i}^r q_{2i} - q_{2i}^r q_{1i} + q_{3i}^r q_{0i} \end{bmatrix} \tag{4.10}$$

where q_i^r denotes desired attitude. q_i^r can be obtained by

$$\begin{bmatrix} q_{0i}^r \\ q_{1i}^r \\ q_{2i}^r \\ q_{3i}^r \end{bmatrix} = \begin{bmatrix} \cos(\phi_i^r/2)\cos(\theta_i^r/2)\cos(\psi_i^r/2) \\ \sin(\phi_i^r/2)\cos(\theta_i^r/2)\cos(\psi_i^r/2) \\ \cos(\phi_i^r/2)\sin(\theta_i^r/2)\cos(\psi_i^r/2) \\ \cos(\phi_i^r/2)\cos(\theta_i^r/2)\sin(\psi_i^r/2) \end{bmatrix} + \begin{bmatrix} \sin(\phi_i^r/2)\sin(\theta_i^r/2)\sin(\psi_i^r/2) \\ -\cos(\phi_i^r/2)\sin(\theta_i^r/2)\sin(\psi_i^r/2) \\ \sin(\phi_i^r/2)\cos(\theta_i^r/2)\sin(\psi_i^r/2) \\ -\sin(\phi_i^r/2)\sin(\theta_i^r/2)\cos(\psi_i^r/2) \end{bmatrix},$$

where ϕ_i^r is the roll angle reference, θ_i^r the pitch angle reference, and ψ_i^r the yaw angle reference. The desired attitude angle references can be given by

$$\phi_i^r = \sin^{-1}(F_{1i}\sin\psi_i^r/u_{0i} - F_{2i}\cos\psi_i^r/u_{0i}),$$

$$\theta_i^r = \tan^{-1}(F_{1i}\cos\psi_i^r/F_{3i} + F_{2i}\sin\psi_i^r/F_{3i}),$$

and $\psi_i^r = 0$, where $u_{0i} = \sqrt{F_{1i}^2 + F_{2i}^2 + F_{3i}^2}$.

Similarly to F_i, the attitude control input $u_{\tau i}$ can be designed with the nominal control part $u_{\tau i}^N$ and the disturbance estimating part $u_{\tau i}^\Delta$ as follows:

$$u_{\tau i} = u_{\tau i}^N + u_{\tau i}^\Delta. \tag{4.11}$$

The nominal control part $u_{\tau i}^N$ is designed as

$$u_{\tau i}^N = \left(\beta_{\tau i}^N\right)^{-1}\left(-K_q e_{qi} - K_\omega e_{\omega i} + (J_i^N)^{-1} S(\omega_i) J_i^N \omega_i + \dot{\omega}_i^r\right), \tag{4.12}$$

where $K_q, K_\omega \in \mathbb{R}^{3\times3}$ are diagonal parameter matrices and the angular velocity ω_i^r can be obtained by $\omega_i^r = 2\,\mathrm{sgn}\left(\sum_{j=0}^{3} q_{ji}^r q_{ji}\right)\tilde{Q}(q_i^r, \dot{q}_i^r)$. Furthermore, similarly to F_i^Δ, $u_{\tau i}^\Delta$ can be constructed as

$$u_{\tau i}^\Delta = -\left(\beta_{\tau i}^N\right)^{-1}\Gamma_{\tau i}(s)\Delta_{Ri}, \tag{4.13}$$

where the robust filter $\Gamma_{\tau i}(s) = diag\{\Gamma_{\tau j,i}(s)\} = diag\{\eta_{\tau j,i}^2 / (s+\eta_{\tau j,i})^2\} \in \mathbb{R}^{3\times3}$ with a positive filter parameter $\eta_{\tau j,i}$. $u_{\tau i}^\Delta$ in (4.13) can be realized similarly to F_i^Δ in (4.9).

Remark 4.4

It can be seen that feedback information of the controllers only depends on its neighbors and itself; thus, the proposed controller subject to switching topologies of each quad-copter is distributed.

4.4 Global System Analysis

The robustness means that all states of the multiple quad-copters control system are bounded, and the tracking errors of each micro quad-copter can converge to a given neighborhood around the origin in a finite time under multiple uncertainties and switching interaction topology. In this section, theoretical analysis is provided to prove the robust stability and tracking performance of the multi-vehicle control system. First, the overall closed-loop error system is constructed to analyze robustness. Then, a theorem is provided based on the small gain theorem to prove the robustness properties of the constructed closed-loop control system.

For $i \in \Phi$, define $e_{1i} = [e_{pTi}^T \quad e_{vTi}^T]^T \in \mathbb{R}^{6\times1}$ and $e_{2i} = [e_{qRi}^T \quad e_{wRi}^T]^T \in \mathbb{R}^{6\times1}$, where $e_{pTi} = [e_{pTj,i}] = p_{Ti} - p_0^r - \delta_i$ and $e_{vTi} = [e_{vTj,i}] = \dot{p}_{Ti} - \dot{p}_0^r$. Combining (4.3), (4.4), (4.5), (4.6), (4.11), and (4.12), one can be obtained that

$$\begin{aligned}
\dot{e}_{1i} &= A_p e_{1i} - \alpha_F B_z \sum_{j\in N_i^{\sigma(t)}} w_t^{ij} K_p(e_{pi} - e_{pj}) - \alpha_F B_z \sum_{j\in N_i^{\sigma(t)}} w_t^{ij} K_v(v_{Ti} - \dot{p}_{Tj}) \\
&\quad - \alpha_F \beta_{li} B_z (K_p e_{pTi} + K_v e_{vTi}) + B_z(\beta_{fi}^N F_i^\Delta + \Delta_{Ti}), \qquad (4.14)\\
\dot{e}_{2i} &= A_q e_{2i} + B_z(\beta_{\tau i}^N u_{\tau i}^\Delta + \Delta_{Ri}),
\end{aligned}$$

where

$$A_p = \begin{bmatrix} 0_{3\times3} & I_3 \\ 0_{3\times3} & 0_{3\times3} \end{bmatrix}, A_q = \begin{bmatrix} 0_{3\times3} & I_3 \\ -K_q & -K_\omega \end{bmatrix}, B_z = \begin{bmatrix} 0_{3\times3} \\ I_3 \end{bmatrix}.$$

Then, the global deviation dynamics of the entire quadrotor group can be attained by the equations:

$$\begin{aligned} \dot{e}_1 &= \left(I_N \otimes A_p - \alpha_F (L_{\sigma(t)} + B_L) \otimes B_z K_T\right) e_1 + \left(I_N \otimes B_z\right) \tilde{\Delta}_T \\ &= A_1^{\sigma(t)} e_1 + B_\Delta \tilde{\Delta}_T, \\ \dot{e}_2 &= \left(I_N \otimes A_q\right) e_2 + \left(I_N \otimes B_z\right) \tilde{\Delta}_R = A_2 e_2 + B_\Delta \tilde{\Delta}_R, \end{aligned} \tag{4.15}$$

where $e_1 = [e_{1i}] \in \mathbb{R}^{6n\times1}$, $e_2 = [e_{2i}] \in \mathbb{R}^{6n\times1}$, $B_L = diag\{\beta_{li}^t\} \in \mathbb{R}^{n\times n}$, $K_T = [K_p \quad K_v]$, $\tilde{\Delta}_T = [\beta_{fi}^N F_i^\Delta + \Delta_{Ti}] \in \mathbb{R}^{3n\times1}$, and $\tilde{\Delta}_R = [\beta_{\tau i}^N u_{\tau i}^\Delta + \Delta_{Ri}] \in \mathbb{R}^{3n\times1}$. The design matrices $Q_T = Q_T^T \in \mathbb{R}^{6\times6}$ and $\Pi_T = \Pi_T^T \in \mathbb{R}^{3\times3}$ are positive definite and symmetric. At $\sigma(t)$, the controller parameter matrix $K_T^{\sigma(t)}$ can be obtained by

$$K_T^{\sigma(t)} = \Pi_T^{-1} B_z^T P_T^{\sigma(t)}, \tag{4.16}$$

where $P_T^{\sigma(t)}$ satisfies

$$(A_1^{\sigma(t)})^T P_T^{\sigma(t)} + P_T^{\sigma(t)} A_1^{\sigma(t)} + Q_T - P_T^{\sigma(t)} B_z \Pi_T^{-1} B_z^T P_T^{\sigma(t)} = 0,$$

and $P_T^{\sigma(t)}$ is the positive definite solution of the above Riccati equation. The root of the directed graph $G_{\sigma(t)}$ can get the relative information from the virtual leader, if $G_{\sigma(t)}$ has a spanning tree. Therefore, at σ(t), it is observed that $A_1^{\sigma(t)T}$ is asymptotically stable if

$$\alpha_F \geq \lambda_{pmR}^{\sigma(t)} / 2, \tag{4.17}$$

where $\lambda_{pmR}^{\sigma(t)} = \min_{i\in\Phi} \mathrm{Re}(\lambda_{pi}^{\sigma(t)})$ and $\lambda_{pi}^{\sigma(t)}$ are the eigenvalues of $(L_{\sigma(t)} + B_L)$. If there are positive diagonal elements in the diagonal matrices K_q and K_ω, one can be observed that A_2 is also asymptotically stable. Let $\lambda_{\sigma(t)}^i (i = 1,2,\ldots,n)$ represent the ith eigenvalue of $L_{\sigma(t)}$. As depicted in [79], the system gets consensus asymptotically subject to directed switching topologies, if the directed graph $G_{\sigma(t)}$ has a spanning tree at $\sigma(t)$ and $K_v = [K_{v1} \quad K_{v2} \quad K_{v3}]^T$ is a constant satisfying $K_{vj} \geq \overline{K}_{vj}$, where $\overline{K}_{vj} \triangleq 0$ $(j = 1,2,3)$, if the $n-1$ nonzero eigenvalues of $L_{\sigma(t)}$ are negative, and

$$\overline{K}_{vj} = \max_{\forall \mathrm{Re}\left(\lambda_{\sigma(t)}^i\right)<0 \text{ and } \mathrm{Im}\left(\lambda_{\sigma(t)}^i\right)>0} \left[\sqrt{2}\left|\lambda_{\sigma(t)}^i\right|^{-1/2} \cos\left(\tan^{-1}\left(\mathrm{Im}(\lambda_{\sigma(t)}^i) / \mathrm{Re}(\lambda_{\sigma(t)}^i)\right)\right)^{-1/2}\right], \tag{4.18}$$

otherwise. The dwell time τ satisfies $\tau > \sup\{D_{\sigma(t)} / W_{\sigma(t)}\}$ at $\sigma(t)$.

Theorem 4.1

Consider the quadrotor dynamics by (4.3) and (4.4), the proposed formation control protocol in the third section. If the directed graph $G_{\sigma(t)}$ always has a spanning tree, its root can get the information including position and velocity from the virtual leader, and the initial states $e_1(0)$ and $e_2(0)$ are bounded. For a given positive constant ε_e, there exist the finite positive constants η_{p1}^* and $\eta_{\tau1}^*$, such that if $\eta_p = \min_{i,j}\{\eta_{pj,i}\} \geq \eta_{p1}^*$ and $\eta_\tau = \min_{i,j}\{\eta_{\tau j,i}\} \geq \eta_{\tau1}^*$ $(i \in \Phi)$, all states in the multi-quadrotors control system are bounded and e_1 and e_2 satisfy $\max_i |e_{1i}(t)| \leq \varepsilon_e$ and $\max_i |e_{2i}(t)| \leq \varepsilon_e, \forall t \geq T^*$.

Proof 4.1 From (4.7), (4.13), and (4.15), one can have that

$$\begin{aligned} \|e_1\|_\infty &\leq \|\pi_{p0}\|_\infty + \gamma_{Bp}\|\Delta_T\|_\infty, \\ \|e_2\|_\infty &\leq \|\pi_{q0}\|_\infty + \gamma_{Bq}\|\Delta_R\|_\infty, \end{aligned} \tag{4.19}$$

where $\pi_{p0} = e^{A_1^{\sigma(t)}t}e_1(0), \pi_{q0} = e^{A_2 t}e_2(0), \Delta_T = [\Delta_{Ti}] \in \mathbb{R}^{3N\times1}, \Delta_R = [\Delta_{Ri}] \in \mathbb{R}^{3N\times1}$, and γ_{Bp}, γ_{Bq} satisfy

$$\begin{aligned} \gamma_{Bp} &= \left\|(sI_{6N} - A_1)^{-1}B_\Delta(I_{3N} - \Gamma_p(s))\right\|_1, \\ \gamma_{Bq} &= \left\|(sI_{6N} - A_2)^{-1}B_\Delta(I_{3N} - \Gamma_\tau(s))\right\|_1, \end{aligned} \tag{4.20}$$

$\Gamma_p(s) = diag\{\Gamma_{pi}(s)\} \in \mathbb{R}^{3n\times3n}$, and $\Gamma_\tau(s) = diag\{\Gamma_{\tau i}(s)\} \in \mathbb{R}^{3n\times3n}$. Because $A_1^{\sigma(t)}$ and A_2 are asymptotically stable and the robust filter would get closer to 1 with the increase of the positive robust filter parameters η_{pi} and $\eta_{\tau i}$ $(i \in \Phi)$, γ_{Bp} and γ_{Bq} can be made as small as desired by selecting the appropriate parameters of the robust filter. Define $\pi_{e(0)} = \max\{\pi_{p0}, \pi_{q0}\}$, $\Delta = [\Delta_T^T \quad \Delta_R^T]^T$, $\gamma_\Delta = \max\{\gamma_{Bp}, \gamma_{Bq}\}$, and $e = [e_1^T \quad e_2^T]^T = [e_k] \in \mathbb{R}^{12n\times1}$. There exist positive constants $\pi_{\Delta T}$ and $\pi_{\Delta R}$ satisfying

$$\|\Delta\|_\infty \leq \pi_{\Delta T}\|e\|_\infty + \pi_{\Delta R}. \tag{4.21}$$

By selecting the disturbance estimator parameters η_{pi} and $\eta_{\tau i}$ $(i \in \Phi)$, one can have that $\gamma_\Delta < \pi_{\Delta I}^{-1}$. Then, combining (4.20) and (4.21), one can obtain that

$$\|e\|_\infty \leq \frac{\pi_{e(0)} + \gamma_\Delta \pi_{\Delta R}}{1 - \gamma_\Delta \pi_{\Delta T}}. \tag{4.22}$$

Since $A_1^{\sigma(t)}$ and A_2 are asymptotically stable and $e(0)$ is bounded, it can be observed that e and Δ are bounded from (4.21) satisfying

$$\|\Delta\|_\infty \leq \pi_\Delta, \tag{4.23}$$

where π_Δ is a positive constant. Besides, it can be seen that the position control inputs F_i and the attitude control inputs $u_{\tau i}$ $(i \in \Phi)$ and the states associated with $u_{\tau i}^\Delta, F_i^\Delta$ are bounded. Therefore, all states of the closed-loop control system are bounded. Substituting (4.7), (4.13), and (4.15) into (4.23), one can obtain that

$$\max_k |e_k(t)| \leq \max_k \left| c_{12N,k}^T e^{At} e(0) \right| + \gamma_\Delta \|\Delta\|_\infty, \tag{4.24}$$

where $A = diag\{A_1^{\sigma(t)}, A_2\}$. There exist positive constants η_{p1}^* and $\eta_{\tau 1}^*$, such that for $\eta_p \geq \eta_{p1}^*$ and $\eta_\tau \geq \eta_{\tau 1}^*$ $(i \in \Phi)$, e_1 and e_2 are bounded and satisfy $\max_i |e_{1i}(t)| \leq \varepsilon_e$ and $\max_i |e_{2i}(t)| \leq \varepsilon_e, \forall t \geq T^*$, respectively.

Remark 4.5

It should be pointed out that the filter parameters η_{pi} and $\eta_{\tau i}$ $(i \in \Phi)$ determined by Theorem 4.1 are conservative, which means that the actual values may be much smaller than their theoretical values. Therefore, in practical applications, one can have the filter parameters using an online tuning method: first, set η_{pi} and $\eta_{\tau i}$ with initial small positive values; second, run the global closed-loop control system; and third, increase η_{pi} and $\eta_{\tau i}$ until the desired tracking performance can be achieved.

4.5 Simulation Results

This section will show the effectiveness of the proposed robust formation control method with switching topologies for a group of four quadrotors, whose dynamics is subject to underactuation, nonlinearities, and disturbances. The four quadrotors are modeled as shown in Section 4.2 with $g = 9.81\ kg/s^2$, $\beta_{\tau i}^N = diag\{9.2, 9.7, 15.99\}$, $J_i^N = diag\{0.109, 0.103, 0.0625\}\ kg \cdot m^2$, and $\beta_{fi}^N = diag\{1,1,1\}$ $(i = 1,2,3,4)$. The quadrotor nominal controller parameters are chosen as $Q_T = diag\{0.06, 0.06, 0.06, 0.1, 0.1, 0.1\}$, $\Pi_T = I_3$, $Q_R = 1000{*}diag\{2,2,2,1,1,1\}$, $\Pi_R = I_3$, and $\alpha_F = 3$.

The directed graph considered here is switching in formation control of the four quadrotors. During $t \in [0, 30s]$, the topologies of the interactions between the four quadrotors are set as $G_{\sigma(0)}$, $G_{\sigma(10)}$, and $G_{\sigma(20)}$, as shown in Figure 4.1. The weighted adjacency matrix $W_{\sigma(t)} = [w_{\sigma(t)}^{ij}] \in \mathbb{R}^{n \times n}$ satisfies that

$$\begin{aligned} W_{\sigma(t)} &= [\varepsilon(t) - \varepsilon(t-10)]W_1 + [\varepsilon(t-10) - \varepsilon(t-20)]W_2 \\ &\quad + [\varepsilon(t-20) - \varepsilon(t-30)]W_3, \end{aligned}$$

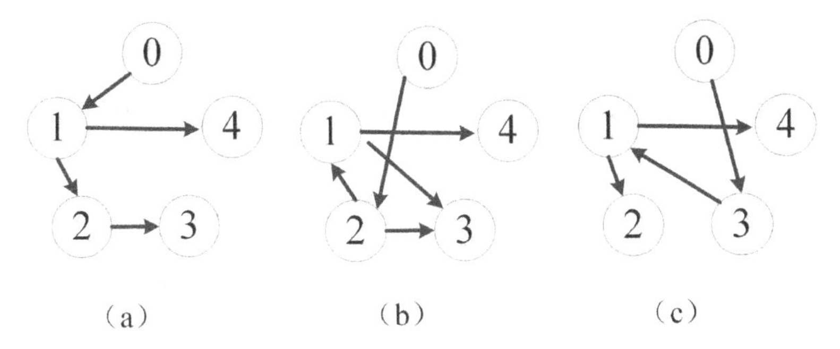

FIGURE 4.1
Switching-directed interaction topologies. (a) Interaction topology $G_{\sigma(0)}$ (b) Interaction topology $G_{\sigma(10)}$ (c) Interaction topology $G_{\sigma(20)}$

where $\varepsilon(t)$ is a step function satisfying $\varepsilon(t)=1$, if $t \geq 0$ and otherwise $\varepsilon(t)=0$, and

$$W_1 = \begin{bmatrix} 0 & 1 & 0 & 1 \\ 0 & 0 & 1 & 0 \\ 0 & 0 & 0 & 0 \\ 0 & 0 & 0 & 0 \end{bmatrix}, W_2 = \begin{bmatrix} 0 & 0 & 1 & 1 \\ 1 & 0 & 0 & 0 \\ 0 & 0 & 0 & 0 \\ 0 & 0 & 0 & 0 \end{bmatrix}, W_3 = \begin{bmatrix} 0 & 1 & 0 & 1 \\ 0 & 0 & 0 & 0 \\ 1 & 0 & 0 & 0 \\ 0 & 0 & 0 & 0 \end{bmatrix}.$$

The time-varying connection weight between the virtual leader and quadrotor i is given by

$$\begin{aligned} \beta_l^t &= [\beta_{l1}^t \quad \beta_{l2}^t \quad \beta_{l3}^t \quad \beta_{l4}^t] \\ &= [\varepsilon(t)-\varepsilon(t-10) \quad \varepsilon(t-10)-\varepsilon(t-20) \quad \varepsilon(t-20)-\varepsilon(t-30) \quad 0]. \end{aligned}$$

The formation center reference of the multi-UAV system is given by $p_0^r = [3t-0.01t^2 \quad 2t-0.02t^2 \quad 3t]^T$. The four quadrotors are required to maintain a constant quadrilateral formation pattern as $\delta_1 = [1 \quad 1 \quad 0]^T$, $\delta_2 = [-1 \quad 1 \quad 0]^T$, $\delta_3 = [1 \quad -1 \quad 0]^T$, and $\delta_4 = [-1 \quad -1 \quad 0]^T$. The actual vehicle parameters are assumed to be 20% larger than the nominal parameters and the time-varying and external disturbance are given by

$$d_{T,fi} = (-1)^i [6\sin(2t) \quad 6\cos(2t) \quad 6\sin(2t)]^T,$$

$$d_{R,\tau i} = (-1)^i [0.3\cos(t) \quad 0.3\sin(t) \quad 0.3\cos(t)]^T.$$

The initial positions and linear velocities for the four quadrotors are selected as $\eta_{l1} = [0.5 \quad 1 \quad 0 \quad 0.2 \quad 0.3 \quad 0.4]^T$, $\eta_{l2} = [-1 \quad 1.5 \quad 0 \quad 0.4 \quad -0.2 \quad 0.2]^T$, $\eta_{l3} = [1 \quad -1.5 \quad 0.5 \quad -0.3 \quad -0.4 \quad -0.3]^T$, and

$\eta_{I4} = [-0.5 \quad -1.5 \quad -1.5 \quad -0.2 \quad 0.4 \quad -0.3]^T$. The initial attitude and the initial angular velocity are set as $q_i(0) = [1 \quad 0 \quad 0 \quad 0]^T$ and $\omega_{Bi}(0) = 0_{3\times1}$.

The trajectory tracking errors using the small disturbance estimating controller parameters with $\eta_{pj,i} = diag\{0.17, 0.5, 0.37\}$ and $\eta_{\tau j,i} = diag\{10, 10, 10\}$ are shown in Figure 4.2. In the results, the solid line, hidden line, dotted line, and chain line represent quadrotor 1, quadrotor 2, quadrotor 3, and quadrotor 4, respectively. Furthermore, increase the values of the disturbance estimating controller parameters to $\eta_{pj,i} = diag\{2.9, 2, 2\}$ and $\eta_{\tau j,i} = diag\{200, 200, 200\}$. The 3D trajectories of the four quadrotors are depicted in Figure 4.3. The linear velocity $(v_{Ixi}, v_{Iyi}, v_{Izi})$ is depicted in Figure 4.4. The attitude responses based on the unit quaternions are depicted in Figure 4.5. The position deviations and the Euler angles of the four quadrotors are illustrated in Figures 4.6 and 4.7, respectively. From these figures, one can observe that the tracking performance is improved by applying larger disturbance estimating controller parameters, especially for the vertical position channel.

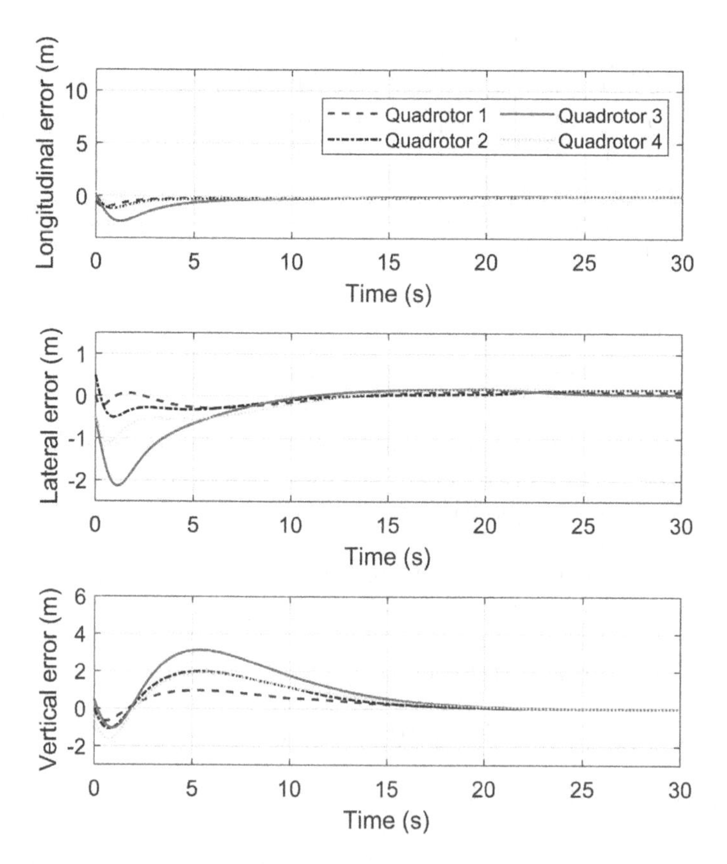

FIGURE 4.2
Position deviations by initial small disturbance estimating controller parameters.

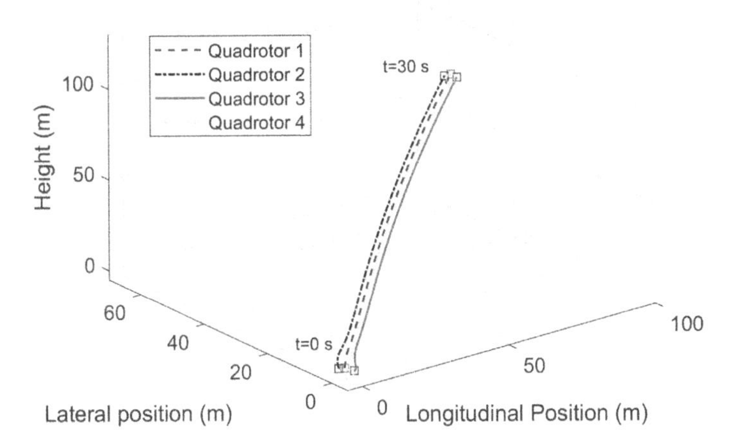

FIGURE 4.3
3D trajectories by larger disturbance estimating controller parameters.

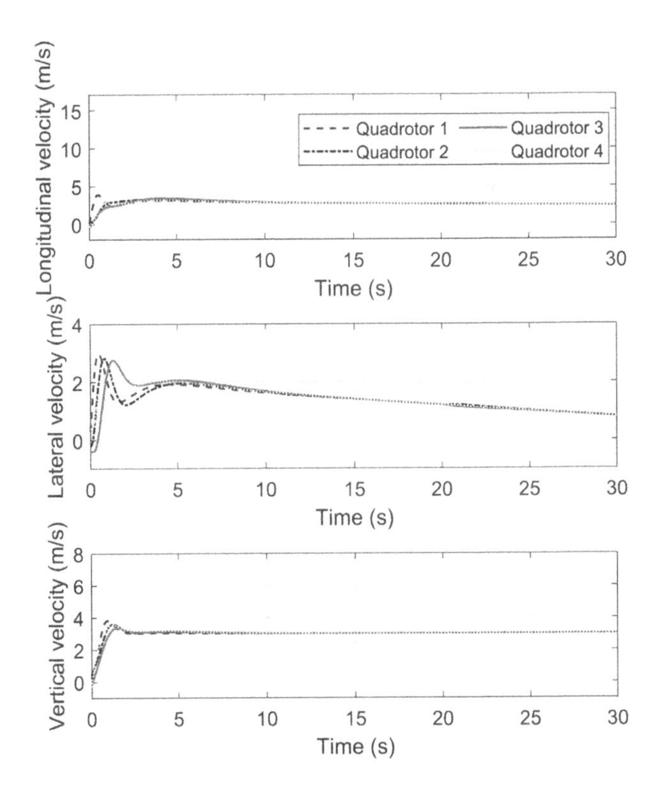

FIGURE 4.4
Linear velocity by a larger disturbance estimating controller.

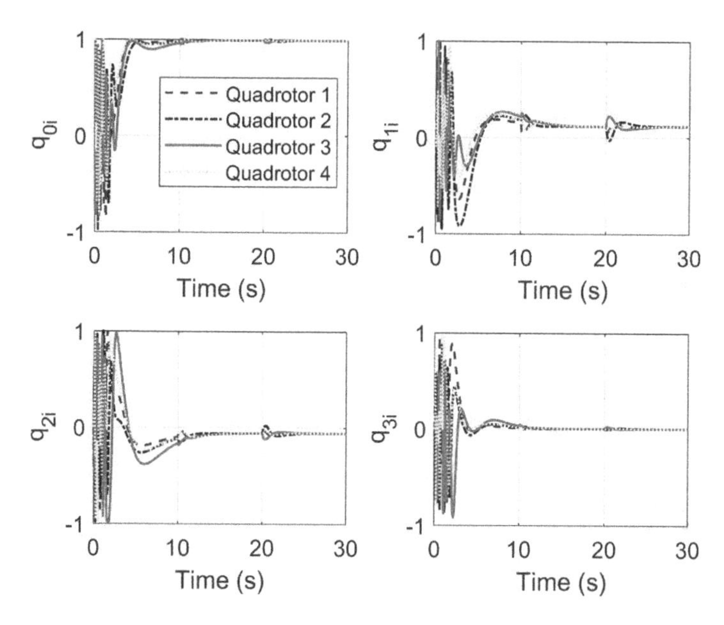

FIGURE 4.5
Attitude responses based on quaternion by larger disturbance estimating controller parameters.

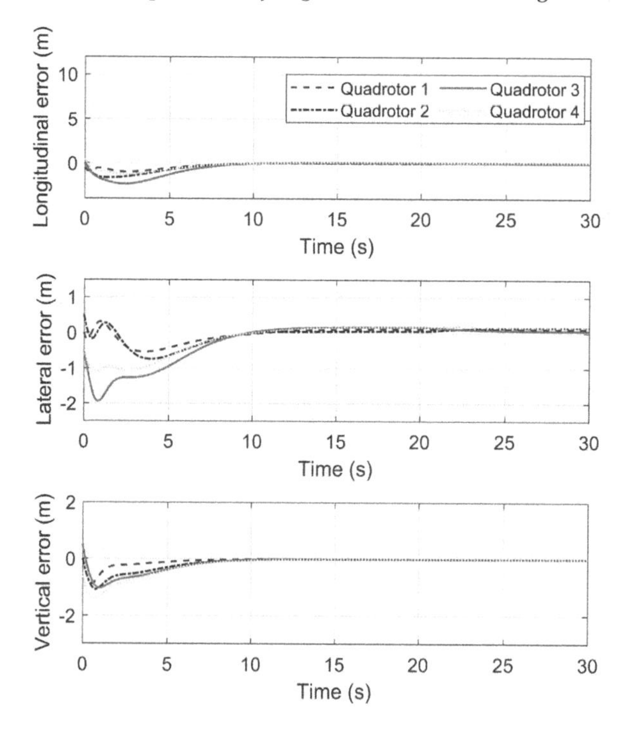

FIGURE 4.6
Position deviations by larger disturbance estimating controller parameters.

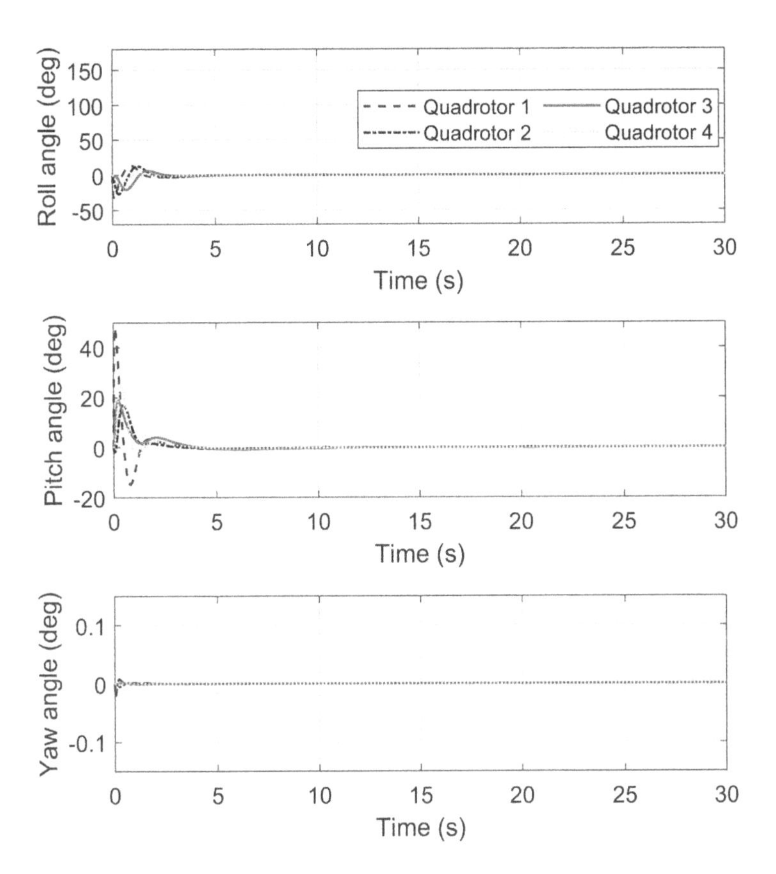

FIGURE 4.7
Euler angle responses by larger disturbance estimating controller parameters.

In contrast, the proposed robust formation controller is compared to a leader-following formation controller developed in [70]. The trajectory tracking deviations by the leader-following method are illustrated in Figure 4.8. The simulation results show that *better* tracking performance and robustness for the multi-UAV system subject to the uncertainties, external disturbance, and switching topologies can be achieved by using the proposed formation protocol.

4.6 Conclusion

In this chapter, the robust time-varying formation control problem is addressed for a set of quadrotors subject to switching topology, whose dynamics involves nonlinearity, parameter perturbations, and

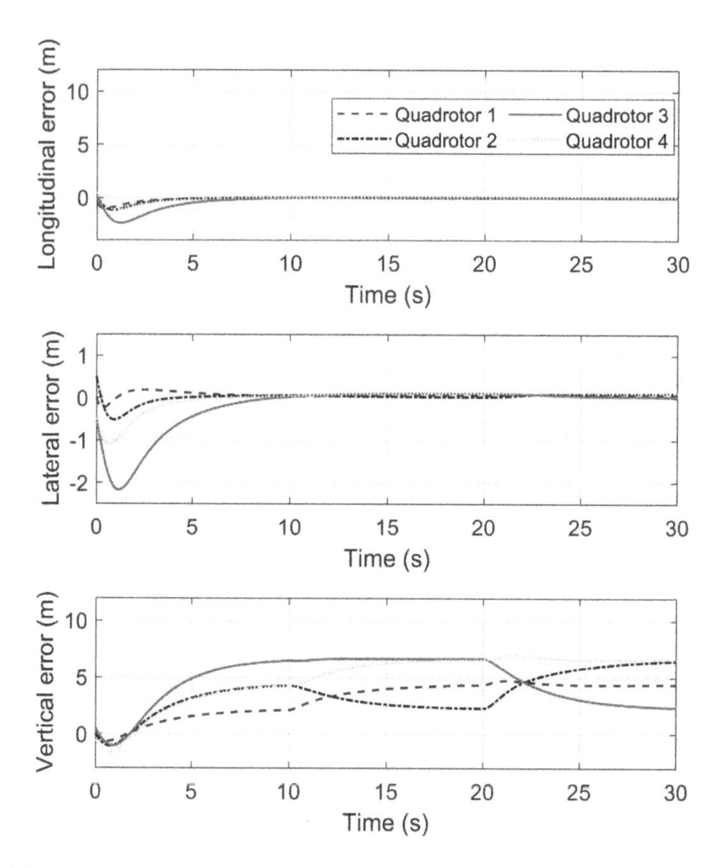

FIGURE 4.8
Position deviations by the leader-following method.

external disturbances. A distributed robust controller is developed consisting of a position controller and an attitude controller for each quadrotor. Both the position controller and the attitude controller involve the nominal controller achieving the desired tracking for the nominal system and the disturbance estimating controller to restrain the influence of uncertainties on the real system. Robustness properties are analyzed and the tracking errors of the formation control system are proven to converge into a given neighborhood of the origin ultimately. The simulation results show the effectiveness of the proposed formation control law.

5

Robust Time-Varying Formation Control for Tail-Sitters in Flight Mode Transitions

This chapter focuses on the problem of the robust formation control for a group of tail-sitters in flight mode transitions. Each tail-sitter dynamics exhibits the features of high nonlinearities and couplings and disturbances in both the translational and rotational motions. A robust distributed formation control method is proposed to achieve aggressive time-varying formation. For each tail-sitter, the proposed control method results in a composite controller that includes a trajectory tracking and an attitude controller to govern the translational and rotational motions, respectively. Theoretical analysis and simulation studies of the formation of multiple tail-sitters are presented to validate the effectiveness of the proposed formation control scheme.

5.1 Introduction

The formation control problem for a group of unmanned aerial vehicle (UAV) systems has attracted significant attention recently in multiple fields [9, 27, 80–82]. The usage of multiple UAVs can reduce the size and complexity required for each single aerial platform and hence lead to a more efficient and effective process. With the increasing demand for multi-UAV systems to cooperate and complete complex tasks, it is important to realize flight formation in a robust manner for these multi-agent systems. However, formation flights present challenges including coordination and cooperation among vehicles, information interaction, communication protocols, and formation controller design.

Tail-sitter UAVs have a number of advantages compared to other configurations [83–86]. In comparison to conventional designs they pose much greater operational flexibility because they don't require a runway for launch and recovery but instead can operate from any small clear space. Although rotary-wing UAVs share the same operational flexibility as the tail-sitter, they suffer from well-known deficiencies in terms of range, endurance, and forward speed limitations due to the lower efficiency of rotor. By marrying the takeoff and landing capabilities of the helicopter with the forward flight efficiencies of fixed-wing aircraft in such a simple way, the tail-sitter

DOI: 10.1201/9781003242147-5

promises a unique blend of capabilities at a lower cost than other UAV configurations. Previously published studies on tail-sitters [87–96] focused on attitude control or trajectory tracking control for a single tail-sitter. However, the formation control problem for multiple tail-sitters is still open. In practical application, different formation patterns can be formed in vertical or horizontal flight using tail-sitters. When a group of tail-sitters takes off vertically, the formation pattern depends on the site conditions. In horizontal flight, a specific formation pattern can reduce air resistance and, thus, save energy. Furthermore, tail-sitters can enter the cruising state with the desired formation pattern after the flight mode transitions as soon as possible.

While the tail-sitter formation concept has great promise, it also comes with significant challenges. First, the flight mode transition of the tail-sitter is an aggressive flight phase between low-speed vertical flight and high-speed forward flight. Each tail-sitter is a multi-variable and under actuated complex system, involving parametric uncertainties, nonlinear and coupled dynamics, and external disturbances. Moreover, a group of tail-sitters is required to achieve different formations in flight mode transition, and thereby the formation is time-varying. Therefore, the aggressive formation control problem for multiple tail-sitters in the flight mode transitions is still open.

Recently, different approaches have been developed for traditional multiple vehicles to achieve formation flight control. In [38], an optimal controller was constructed for multiple classical fixed-wing UAVs. A robust formation control strategy was proposed in [44] for multiple helicopters to achieve the desired formation pattern. In [55], a distributed feedback formation controller was designed for a group of UAVs to achieve time-invariant formation flight. A model predictive formation control approach was presented in [99] for traditional quadcopters with nonlinear models to implement time-invariant formation tracking missions. In [100], a composite nonlinear feedback controller based on the leader-follower structure was developed for a group of typical quadrotors to keep a specified formation pattern.

This chapter mainly investigates the formation flight control problem for the novel tail-sitter UAVs developed to achieve the desired time-varying formation pattern in the flight mode transitions. A robust formation control method is proposed for a team of tail-sitters. Compared with the previous relevant results, the contributions of this chapter are threefold.

First, multiple tail-sitters can achieve the time-varying formation pattern in the flight mode transitions. Second, aggressive continuous flight mode transitions between the low-speed vertical flight and the high-speed forward flight can be achieved for a group of tail-sitters. The proposed control method does not require switching on the coordinate system or the controller parameters in different flight modes. Third, for the dynamical model of each tail-sitter, nonlinear and coupled dynamics, parametric uncertainties, and external disturbances are considered. The robustness properties of the constructed control system can be guaranteed and the tracking errors of the global system can converge to a neighborhood of the origin in a finite time.

5.2 Preliminaries and Problem Statement

5.2.1 Model of Tail-Sitter Aircraft

The schematic of a typical tail-sitter aircraft, as illustrated in [84], is shown in Figure 5.1. The tail-sitter UAV has a unique fixed-wing quadcopter hybrid design consisting of a fuselage, four rotors, coaxial counter-rotating propellers, two wings, and two ailerons. One of the main advantages of this configuration is that the quadrotors can control the attitude in both hover and transition flight modes. A typical flight process of the tail-sitter can consist of four phases: hover, hover to forward flight, forward flight, and forward flight to hover as shown in Figure 5.2. During vertical flight or hover, the model is similar to standard quadrotor aircraft. The main lift force is generated by the coaxial counter-rotating propellers, and the control torques are provided by the four rotors to stabilize the attitude angles. In the forward flight, the lift force is generated by the wings. Besides, the dominant thrust is provided by the coaxial counter-rotating propellers. The pitch and yaw dynamics is achieved by controlling Rotors 1–4 and the roll channel is controlled by two ailerons to save energy.

When the vehicle is in the transition flight, the control torques are generated by the differential force of the two pairs of rotors and the two ailerons.

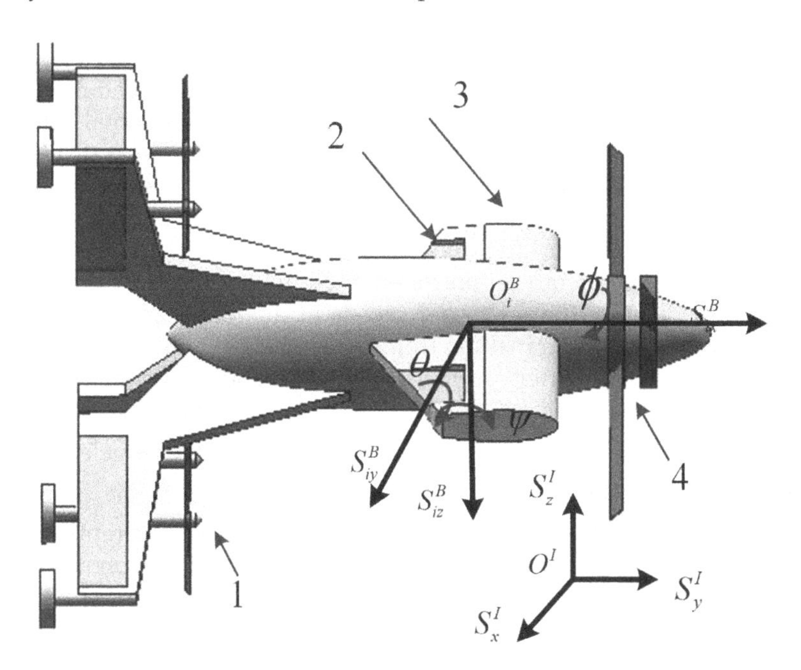

FIGURE 5.1
Schematic of the tail-sitter aircraft. (1: Rotors; 2: ailerons; 3: wings; 4: coaxial counter-rotating propellers.)

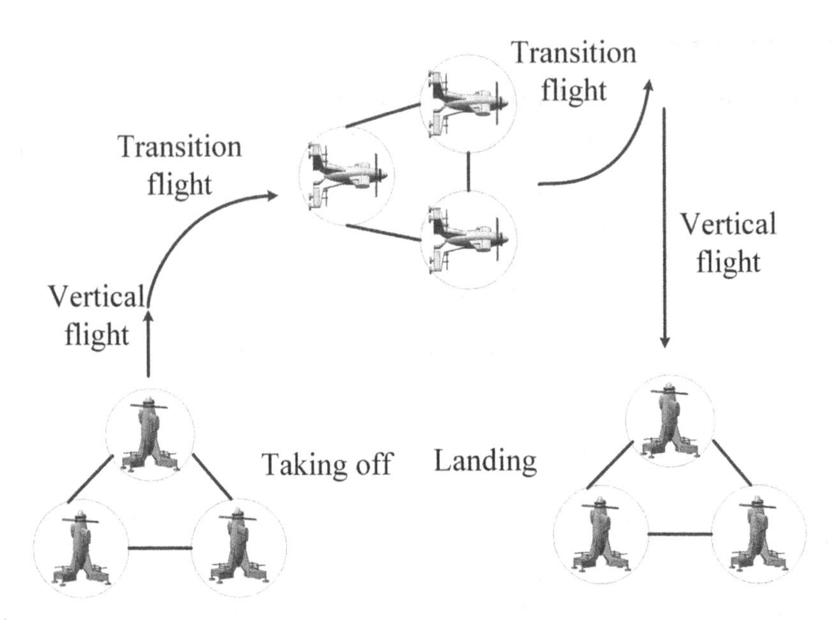

FIGURE 5.2
Formation schematics for a group of tail-sitters.

As depicted in Figure 5.2, from hover to forward flight, the forces generated by Rotors 3 and 4 increase gradually, while the forces generated by Rotors 1 and 2 decrease. Thus, the fuselage is sloped toward the horizontal, where the pitch angle is nearly 0°. Therefore, the flight speed of the aircraft increases gradually, and an angle of attack of the fixed wings would be obtained to generate a lift force to support the entire body weight. Then, the flight mode is switched to the forward flight mode to complete transition. From forward flight to hover, the aircraft is controlled to climb up and reduce the aircraft velocity by adjusting the rotational speeds of the four rotors. Thereby, the aircraft implements the transition from cruising flight to hover.

Consider the tail-sitter formation coordination system with two frames: the inertial frame $E^I = \{O^I, E_x^I, E_y^I, E_z^I\}$ and the body-fixed frame $E_i^B = \{O_i^B, E_{ix}^B, E_{iy}^B, E_{iz}^B\}$, where O_i^B is fixed at the center of the vehicle mass. The attitude vector is denoted as $\Phi_i = \begin{bmatrix} \phi_i & \theta_i & \psi_i \end{bmatrix}^T$, where ϕ_i is the roll angle, θ_i is the pitch angle, and ψ_i is the yaw angle. Because the Euler representation suffers from the singularity problem, the unit quaternion representation is applied to describe the rotational dynamics in the flight mode transitions. We define $q_i = \begin{bmatrix} q_{i0} & \bar{q}_i \end{bmatrix}^T$, where $\bar{q}_i = \begin{bmatrix} q_{i1} & q_{i2} & q_{i3} \end{bmatrix}^T$ and q_{i0} are the vector part and the scalar part of the quaternion, respectively, and satisfy $\bar{q}_i^T \bar{q}_i + q_{i0}^2 = 1$. The coordinate rotation matrix from S_i^B to S^I described by the quaternions can be given by

$$R_i = \begin{bmatrix} 1-2q_{1i}^2-2q_{3i}^2 & 2q_{1i}q_{2i}-2q_{0i}q_{3i} & 2q_{1i}q_{3i}+2q_{0i}q_{2i} \\ 2q_{1i}q_{2i}+2q_{0i}q_{3i} & 1-2q_{1i}^2-2q_{3i}^2 & 2q_{2i}q_{3i}-2q_{0i}q_{1i} \\ 2q_{1i}q_{3i}-2q_{0i}q_{2i} & 2q_{2i}q_{3i}+2q_{0i}q_{1i} & 1-2q_{1i}^2-2q_{2i}^2 \end{bmatrix}.$$

For tail-sitter i, we define $p_{Ii} = \begin{bmatrix} p_{Ixi} & p_{Iyi} & p_{Izi} \end{bmatrix}^T$ as the position of the center of gravity in S^I, where p_{Ixi} represents the longitudinal position, p_{Iyi} is the lateral position, and p_{Izi} is the vertical position. Let $J_i = diag\left(J_{xi}, J_{yi}, J_{zi}\right)$ be the inertia matrix of the tail-sitter and m_i be the mass of the tail-sitter. We denote $v_{Bi} = \begin{bmatrix} v_{Bxi} & v_{Byi} & v_{Bzi} \end{bmatrix}^T$ as the velocity of tail-sitter i relative to S_i^B and $\omega_{Bi} = \begin{bmatrix} \omega_{Bxi} & \omega_{Byi} & \omega_{Bzi} \end{bmatrix}^T$ the angular speed. We define g as the acceleration of gravity and m_i the mass of tail-sitter i. Let the superscript $\times$ represent the cross product and $c_{j,n}$ an $n \times 1$ column vector with one on the j-th row and zeros elsewhere for tail-sitter i. From [84], the dynamical model of tail-sitter i can be given as

$$\begin{aligned} m_i \ddot{p}_{Ii} &= R_i\left(F_{ti} + F_{wi} + F_{fi} + F_{di}\right) - m_i g c_{3,3}, \\ J_i \dot{\omega}_{Bi} + \omega_{Bi}^{\times} J_i \omega_{Bi} &= \tau_{ci} + \tau_{wi} + \tau_{gi} + \tau_{di}, \\ \dot{\bar{q}}_i &= 0.5\bar{q}_i^{\times}\omega_{Bi} + 0.5 q_{0i}\omega_{Bi}, \\ \dot{q}_{0i} &= -0.5\omega_{Bi}^T \bar{q}_i, \end{aligned} \tag{5.1}$$

where F_{ti} denotes the total thrust produced by the coaxial counter-rotating propellers and the four rotors, F_{wi} and τ_{wi} denote the aerodynamic force and moment by the two fixed wings, respectively, F_{fi} denotes the aerodynamic force by the fuselage, τ_{ci} represents the control torque generated by the four rotors and the two ailerons, τ_{gi} is the moment generated by the gyroscopic effects, and F_{di} and τ_{di} denote the external disturbance force and moment, respectively. The moment τ_{ci} can be calculated by

$$\tau_{ci} = \tau_{ri} + \tau_{ai}, \tag{5.2}$$

where τ_{ri} is the moment generated by the four rotors and τ_{ai} is the moment by the two ailerons. F_{ti} can be obtained by $F_{ti} = F_{ri} + F_{ci}$, where F_{ri} is the total thrust generated by the four rotors and F_{ci} the thrust by the coaxial counter-rotating propellers. The forces F_{ri} and F_{ci} acting on the tail-sitter aircraft can be given as follows:

$$F_{ri} = \left[k_{ri} \sum_{j=1}^{4} \omega_{ji}^2 \quad 0 \quad 0 \right]^T, \quad F_{ci} = \left[k_{ci}\omega_{ci}^2 \quad 0 \quad 0 \right]^T, \tag{5.3}$$

where k_{ri} is the thrust coefficient for each rotor, $\omega_{ji}\left(j=1,\ 2,3,4\right)$ represent the rotational velocities of Rotors 1–4, respectively, k_{ci} is the thrust coefficient for the coaxial counter-rotating propellers, and ω_{ci} is the rotational speed of the coaxial counter-rotating propellers. The thrusts F_{ri} and F_{ci} satisfy the mathematical relationship that $F_{ci}=k_{fi}F_{ri}$, where k_{fi} is a positive constant. F_{wi} and F_{fi} can be expressed as

$$F_{wi}=\begin{bmatrix}(L_{1i}+L_{2i})\sin\alpha_i-(D_{1i}+D_{2i})\cos\alpha_i\\0\\-(L_{1i}+L_{2i})\cos\alpha_i-(D_{1i}+D_{2i})\sin\alpha_i\end{bmatrix},\ F_{fi}=\begin{bmatrix}L_{fi}\sin\alpha_i-D_{fi}\cos\alpha_i\\0\\-L_{fi}\cos\alpha_i-D_{fi}\sin\alpha_i\end{bmatrix},\tag{5.4}$$

where L_{1i} and L_{2i} are the lift forces, D_{1i} and D_{2i} are the drag forces, and L_{fi} and D_{fi} are the lift and drag forces generated by the fuselage, respectively. The lift and drag forces acting on the tail-sitters can be calculated as

$$\begin{aligned}L_{ji}&=0.5C_{Lji}S_{ri}\rho\left(v_{Bxi}^2+v_{Bzi}^2\right),\\D_{ji}&=0.5C_{Dji}S_{ri}\rho\left(v_{Bxi}^2+v_{Bzi}^2\right),\\L_{fi}&=0.5C_{lfi}S_{fi}\rho\left(v_{Bxi}^2+v_{Bzi}^2\right),\\D_{fi}&=0.5C_{dfi}S_{fi}\rho\left(v_{Bxi}^2+v_{Bzi}^2\right),\ j=1,\ 2,\end{aligned}$$

where S_{ri} is the reference area, ρ is the air density, S_{fi} is the blade area, and C_{L1i}, C_{L2i}, C_{D1i}, C_{D2i}, C_{lfi}, and C_{dfi} are the thrust or drag coefficients. These aerodynamic coefficients are related to α_i and can be obtained by

$$\begin{aligned}C_{Lji}&=C_{L0i}+C_{L\alpha i}\,\alpha_i+C_{L\delta i}\,\sigma_{ji},\\C_{Dji}&=C_{D0i}+C_{Lji}^2/a_{ei},\\C_{lfi}&=C_{lf\alpha i}\alpha_i,\\C_{dfi}&=C_{df0i}+C_{df\alpha i}\,\alpha_i,\ j=1,\ 2,\end{aligned}$$

where C_{L0i}, $C_{L\alpha i}$, $C_{L\sigma i}$, C_{D0i}, $C_{lf\alpha i}$, C_{df0i}, and $C_{df\alpha i}$ are the aerodynamic coefficients, σ_{1i} and σ_{2i} are the flap bias angles, and a_{ei} is a positive constant. Moments τ_{ri} and τ_{wi} acting on the tail-sitter aircraft can be obtained as follows:

$$\tau_{ri}=\begin{bmatrix}\sum_{j=1}^{4}(-1)^{j+1}\left(a_{\tau 1i}+\sqrt{2}a_{\tau 2i}l_{\tau 1i}/2\right)\omega_{ji}^2\\l_{\tau 1i}\left[F_{r1i}+F_{r2i}-F_{r3i}-F_{r4i}\right]/2\\l_{\tau 1i}\left[-F_{r1i}+F_{r2i}+F_{r3i}-F_{r4i}\right]/2\end{bmatrix},\tag{5.5}$$

and

$$\tau_{wi} = \begin{bmatrix} l_{\tau 2i}(L_{2i} - L_{1i})\cos\alpha_i + l_{\tau 2i}(D_{2i} - D_{1i})\sin\alpha_i \\ l_{\tau 3i}(L_{2i} + L_{1i})\cos\alpha_i + l_{\tau 3i}(D_{2i} + D_{1i})\sin\alpha_i \\ l_{\tau 2i}(L_{2i} - L_{1i})\sin\alpha_i + l_{\tau 2i}(D_{1i} - D_{2i})\cos\alpha_i \end{bmatrix}, \tag{5.6}$$

where $a_{\tau 1i}$ and $a_{\tau 2i}$ are positive constants, $l_{\tau 1i}$ is the distance between two rotors, $l_{\tau 2i}$ is the distance between the center of gravity of the aircraft and the force acting position of the fixed wing, and $l_{\tau 3i}$ is the distance between the center of gravity of the aircraft and the fixed wing. τ_{ai} can be expressed as

$$\tau_{ai} = \begin{bmatrix} l_{\tau 2i} C_{L\sigma i} S_{ri} \rho \cos\alpha_i \left(v_{Bxi}^2 + v_{Bzi}^2\right)(\sigma_{1i} - \sigma_{2i}) \\ l_{\tau 3i} C_{L\sigma i} S_{ri} \rho \cos\alpha_i \left(v_{Bxi}^2 + v_{Bzi}^2\right)(\sigma_{1i} + \sigma_{2i}) \\ 0 \end{bmatrix}. \tag{5.7}$$

τ_{gi} can be obtained by

$$\tau_{gi} = \sum_{j=1}^{4} J_{ri} \omega_{Bi}^{\times} c_{3,3} (-1)^i \omega_{ji},$$

where J_{ri} is the rotating inertia of each rotor.

Remark 5.1

It can be observed that the vehicle system is highly nonlinear and strongly coupled. The coordinate system of the tail-sitter system is consistent in every flight mode.

5.2.2 Control Problem Statement

We define $\delta_{ij} = \begin{bmatrix} \delta_{xij} & \delta_{yij} & \delta_{zij} \end{bmatrix}^T \in \mathbb{R}^{3\times 1}$ as the desired time-varying position deviation between tail-sitter i and tail-sitter j, which determines the time-varying formation pattern of the tail-sitter group. Let $p_0^r = \begin{bmatrix} p_{x0}^r & p_{y0}^r & p_{z0}^r \end{bmatrix}^T \in \mathbb{R}^{3\times 1}$ as the prescribed desired trajectory of a virtual leader, where its second derivative $\ddot{p}_0^r$ is assumed to be bounded and could converge to 0 in a finite time. We denote δ_i as the position deviation between the virtual leader and tail-sitter i and it follows that $\delta_{ij} = \delta_i - \delta_j$. The control goal in the current chapter is to propose a robust distributed control protocol to achieve the desired time-varying patterns and track the desired formation trajectory.

From the first and second equations in (5.1), one can have that

$$\begin{aligned}\ddot{p}_{Ii} &= m_i^{-1}R_i\left(F_{ti}\right)+m_i^{-1}R_i\left(F_{wi}+F_{fi}\right)+m_i^{-1}R_iF_{di}-gc_{3,3},\\ \dot{\omega}_{Bi} &= -J_i^{-1}\omega_{Bi}^{\times}J_i\omega_{Bi}+J_i^{-1}\tau_{ci}+J_i^{-1}\left(\tau_{wi}+\tau_{gi}\right)+J_i^{-1}\tau_{di}.\end{aligned} \tag{5.8}$$

Let $B_{pi}=m_i^{-1}I_3$ and $B_{qi}=J_i^{-1}$, where I_n represents an $n\times n$ unit matrix. We define the superscript N as the nominal parameter and the superscript Δ as the parameter uncertainty. Then, one can have that $B_{pi}=B_{pi}^N+B_{pi}^\Delta$ and $B_{qi}=B_{qi}^N+B_{qi}^\Delta$. We denote the attitude control input $u_{ai}=\tau_{ci}$ and the virtual position control input $u_{pi}=R_iF_{ti}$. The dynamics of tail-sitter i in (5.8) can be rewritten as

$$\begin{aligned}\ddot{p}_{Ii} &= B_{pi}^N u_{pi}+B_{pi}^N R_i\left(F_{wi}+F_{fi}\right)-gc_{3,3}+\Delta_{pi},\\ \dot{\omega}_{Bi} &= -B_{qi}^N\omega_{Bi}^{\times}J_i^N\omega_{Bi}+B_{qi}^N u_{ai}+B_{qi}^N\left(\tau_{wi}+\tau_{gi}\right)+\Delta_{ai},\end{aligned} \tag{5.9}$$

where Δ_{pi}, Δ_{ai} are named the equivalent disturbances including the parametric uncertainties, nonlinear and coupled dynamics, and external disturbances and have the following forms:

$$\begin{aligned}\Delta_{pi} &= B_{pi}^\Delta R_iF_{ti}+B_{pi}^\Delta R_i\left(F_{wi}+F_{fi}\right)+m_i^{-1}R_iF_{di},\\ \Delta_{ai} &= B_{qi}\omega_{Bi}^{\times}J_i^\Delta\omega_{Bi}+B_{qi}^\Delta\omega_{Bi}^{\times}J_i^N\omega_{Bi}+B_{qi}^\Delta u_{ai}+B_{qi}^\Delta\left(\tau_{wi}+\tau_{gi}\right)+\left(J_i^N\right)^{-1}\tau_{di}.\end{aligned} \tag{5.10}$$

Remark 5.2

It should be noted that the tail-sitter system is a complex system subject to multiple actuator faults, highly nonlinear and coupled dynamics, external disturbances, and parametric uncertainties, which pose challenges for the fault-tolerant formation controller design. By ignoring Δ_{ip} and Δ_{ia}, the model in (6.12) represents the nominal model. The real vehicle model can be regarded as the nominal model added with Δ_{ip} and Δ_{ia}.

5.3 Robust Formation Controller Design

In this section, a robust formation controller is designed for the team of tail-sitters to achieve the desired time-varying formation scenario. The overall controller design is divided into two parts: the trajectory tracking controller design to govern the translational dynamics with the desired formation trajectory and pattern and the attitude controller design to stabilize the inner rotational dynamics.

5.3.1 Trajectory Tracking Controller Design

For tail-sitter i, the position control input u_{pi} is designed as

$$u_{pi} = u_{pi}^N + u_{pi}^R. \tag{5.11}$$

where u_{pi}^N denotes the nominal control input and u_{pi}^R denotes the robust compensating input. The nominal controller based on the state feedback control method is designed to achieve the desired tracking performance for the nominal position system by ignoring the equivalent disturbance Δ_{pi}, while the robust compensator is introduced to restrain the effects of Δ_{pi}.

The nominal control input is constructed as

$$\begin{aligned} u_{pi}^N = & -\lambda_s \left(B_{pi}^N\right)^{-1} K_p \left[\sum_{j\in N_i} w_{ij}\left(p_i - p_j - \delta_{ij}\right) + k_{ci}\left(p_i - \delta_i - p_0^r\right)\right] \\ & -\lambda_s \left(B_{pi}^N\right)^{-1} K_{\dot{p}} \left[\sum_{j\in N_i} w_{ij}\left(\dot{p}_i - \dot{p}_j - \dot{\delta}_{ij}\right) + k_{ci}\left(\dot{p}_i - \dot{\delta}_i - \dot{p}_0^r\right)\right] \\ & - R_i\left(F_{wi} + F_{fi}\right) + \left(B_{pi}^N\right)^{-1}\left(gc_{3,3} + \ddot{\delta}_i\right), \end{aligned} \tag{5.12}$$

where λ_s represents the coupling gain, K_p and $K_{\dot{p}}$ are the nominal controller parameter matrices, and k_{ci} is the connection weight between tail-sitter i and the virtual leader. $k_{ci} > 0$ means that tail-sitter i can receive the information from the virtual leader; otherwise, $k_{ci} = 0$.

Moreover, the robust compensating input u_{pi}^R can be constructed by the following form:

$$u_{pi}^R(s) = -\left(B_{pi}^N\right)^{-1} F_{pi}(s)\Delta_{pi}(s), \tag{5.13}$$

where s is the Laplace operator, $F_{pi}(s) = diag\left\{F_{pi1}(s), F_{pi2}(s), F_{pi3}(s)\right\}$, and the robust filters $F_{pij}(s) = f_{pij}^2 / \left(s + f_{pij}\right)^2 \left(j = 1,\ 2,\ 3\right)$. In fact, larger robust compensator parameters $f_{pij}\left(j = 1,\ 2,\ 3\right)$ lead to wider frequency bandwidths of the robust filters $F_{pij}(s)$ $\left(j = 1,\ 2,\ 3\right)$ and the robust filter gains are closer to 1. In this case, the effects of the equivalent disturbance Δ_{pi} would be restrained, as illustrated in [62]. In practical applications, the equivalent disturbance Δ_{pi} cannot be measured directly. Therefore, the robust control input u_{pi}^R does not depend on Δ_{pi}. From (5.9), one can have that

$$\Delta_{pi} = \ddot{p}_{li} - B_{pi}^{N} u_{pi} - B_{pi}^{N} R_i \left(F_{wi} + F_{fi} \right) + g c_{3,3}. \tag{5.14}$$

From (5.13) and (5.14), one can obtain the realization of the robust compensator as follows:

$$\begin{aligned} \dot{\eta}_{p1i} &= -f_{pi}\eta_{p1i} - f_{pi}^{2} p_{li} + B_{pi}^{N} u_{pi} + B_{pi}^{N} R_i \left(F_{wi} + F_{fi} \right) - g c_{3,3}, \\ \dot{\eta}_{p2i} &= -f_{pi}\eta_{p2i} + 2 f_{pi} p_{li} + \eta_{p1i}, \\ u_{pi}^{R} &= \left(B_{pi}^{N} \right)^{-1} f_{pi}^{2} \left(\eta_{p2i} - p_{li} \right), \end{aligned} \tag{5.15}$$

where η_{p1i} and η_{p2i} are filter states and $f_{pi} = diag\left\{ f_{pi1}, f_{pi2}, f_{pi3} \right\}$.

5.3.2 Attitude Controller Design

For tail-sitter i, the attitude reference is denoted as $q_{ri} = \begin{bmatrix} q_{0ri} & q_{1ri} & q_{2ri} & q_{3ri} \end{bmatrix}^T$. We define the attitude tracking error e_{kai} as $e_{kai} = \begin{bmatrix} e_{Qi} & e_{\omega i} \end{bmatrix}^T = \begin{bmatrix} \tilde{Q}_i & \omega_{Bi} - \omega_{Bi}^{r} \end{bmatrix}^T$, where $\tilde{Q}_i$ is the attitude error based on the quaternion representation and ω_{Bi}^{r} is the desired angular velocity. The nonlinear function $\tilde{Q}_i$ can be given as follows:

$$\tilde{Q}\left(q_i, q_{ri} \right) = 2\,\mathrm{sgn}\left(\sum\nolimits_{j=0}^{3} q_{ji} q_{jri} \right) \begin{bmatrix} -q_{0i}q_{1ri} + q_{1i}q_{0ri} + q_{2i}q_{3ri} - q_{3i}q_{2ri} \\ -q_{0i}q_{2ri} - q_{1i}q_{3ri} + q_{2i}q_{0ri} + q_{3i}q_{1ri} \\ -q_{0i}q_{3ri} + q_{1i}q_{2ri} - q_{2i}q_{1ri} + q_{3i}q_{0ri} \end{bmatrix}.$$

The desired angular velocity ω_{Bi}^{r} can be obtained by $\omega_{Bi}^{r} = 0.5\begin{bmatrix} -\bar{q}_{ri} & \left(q_{0ri} I_3 - \bar{q}_{ri}^{\times} \right)^T \end{bmatrix} \dot{q}_{ri}$. The error dynamics of the rotational subsystem can be given as: $\dot{e}_{kai} = \begin{bmatrix} \omega_{Bi} - \omega_{Bi}^{r} & \dot{\omega}_{Bi} - \dot{\omega}_{Bi}^{r} \end{bmatrix}^T$. Similarly to the trajectory tracking control input u_{pi}, the attitude control input u_{ai} also involves the nominal control input u_{ai}^{N} and the robust compensating input u_{ai}^{R} as follows:

$$u_{ai} = u_{ai}^{N} + u_{ai}^{R}. \tag{5.16}$$

u_{ai}^{N} can be designed as follows:

$$u_{ai}^{N} = \left(B_{qi}^{N} \right)^{-1} \left(-K_a e_{Qi} - K_{\dot{a}} e_{wi} + \dot{\omega}_{Bi}^{r} \right) + \omega_{Bi}^{\times} J_i^{N} \omega_{Bi} - \left(\tau_{wi} + \tau_{gi} \right). \tag{5.17}$$

Furthermore, u_{ai}^{R} is constructed as follows:

$$u_{ai}^{R}(s) = -\left(B_{qi}^{N} \right)^{-1} F_{ai}(s) \Delta_{ai}(s), \tag{5.18}$$

where the robust filter satisfies $F_{ai}(s) = diag\{F_{ai1}(s), F_{ai2}(s), F_{ai3}(s)\}$ and $F_{aij}(s) = f_{aij}^2/(s+f_{aij})^2\ (j=1,2,3)$. The implementation of u_{ai}^R is similar to that of u_{pi}^R as follows:

$$\dot{\eta}_{a1i} = -f_{ai}\eta_{a1i} - f_{ai}^2 e_{iQ} + B_{qi}^N u_{ai} - B_{qi}^N \omega_{Bi}^{\times} J_i^N \omega_{Bi} + B_{qi}^N \left(\tau_{wi} + \tau_{gi}\right) - \dot{\omega}_{Bi}^r,$$

$$\dot{\eta}_{a2i} = -f_{ai}\eta_{a2i} + 2f_{ai}e_{iQ} + \eta_{a1i},$$

$$u_{ai}^R = \left(B_{qi}^N\right)^{-1} f_{ai}^2 \left(\eta_{a2i} - e_{iQ}\right),$$

where η_{a1i} and η_{a2i} are filter states and $f_{ai} = diag\{f_{ai1}, f_{ai2}, f_{ai3}\}$

Remark 5.3

The designed controller is continuous in the flight mode transitions. There is no need to switch the controller structures or the controller parameters. Besides, the proposed fault-tolerant time-varying formation control strategy is distributed because the designed control law of each tail-sitter depends on the position and velocity information from its neighbors and itself.

5.4 Robust Property Analysis

Define the trajectory tracking error $e_{pi} = [e_{pi,j}] = p_{li} - \delta_i - p_0^r$ and $e_{kpi} = \left[e_{pi}^T \quad \dot{e}_{pi}^T\right]^T$. From (5.9), (5.11), (5.12), (5.16), and (5.17), one can obtain that

$$\dot{e}_{kpi} = A_{kp}e_{kpi} - \lambda_s B_{k1} K_p \left[\sum_{j\in N_i} w_{ij}\left(e_{pi} - e_{pj}\right) + k_{ci}e_{pi}\right]$$

$$- \lambda_s B_{k1} K_{\dot{p}} \left[\sum_{j\in N_i} w_{ij}\left(e_{vi} - e_{vj}\right) + k_{ci}e_{vi}\right] + B_{k1}\left(B_{pi}^N u_{pi}^R + \Delta_{pi} - \ddot{p}_0^r\right),$$

$$\dot{e}_{kai} = A_{ka}e_{kai} + B_{k1}\left(B_{qi}^N u_{ai}^R + \Delta_{ai}\right), \tag{5.19}$$

where

$$A_{kp} = \begin{bmatrix} 0_{3\times3} & I_3 \\ 0_{3\times3} & 0_{3\times3} \end{bmatrix}, A_{ka} = \begin{bmatrix} 0_{3\times3} & I_3 \\ -K_a & -K_{\dot{a}} \end{bmatrix}, B_{k1} = \begin{bmatrix} 0_{3\times3} \\ I_3 \end{bmatrix},$$

$$A_{ka} = \begin{bmatrix} 0_{3\times3} & I_3 \\ -K_a & -K_{\dot{a}} \end{bmatrix},$$

and

$$B_{k1} = \begin{bmatrix} 0_{3\times3} \\ I_3 \end{bmatrix}.$$

Then, the overall closed-loop error system can be written as

$$\begin{aligned} \dot{e}_p &= A_p e_p + B_{k2}\left(B_{pi}^N u_{pi}^R + \Delta_{pi} - \ddot{p}_0^r\right), \\ \dot{e}_a &= A_a e_a + B_{k2}\left(B_{qi}^N u_{ai}^R + \Delta_{ai}\right), \end{aligned} \tag{5.20}$$

where $e_p = \left[e_{kpi}\right] \in \mathbb{R}^{6N\times1}$, $e_a = \left[e_{kai}\right] \in \mathbb{R}^{6N\times1}$, and

$$\begin{aligned} A_p &= I_N \otimes A_{kp} - \lambda_s\left(L + B_L\right) \otimes B_k K_t, \\ A_q &= I_N \otimes A_{ka},\ B_{k2} = I_N \otimes B_{k1}, \end{aligned}$$

with $B_L = diag\{k_{ci}\} \in \mathbb{R}^{N\times N}$ and $K_t = \left[K_p \quad K_{\dot{p}}\right]$. One can observe that the matrix A_q is asymptotically stable by selecting proper diagonal gain matrices K_a and $K_{\dot{a}}$ with positive diagonal elements. If the directed graph G has a spanning tree, the root can obtain the information from the virtual leader, and $\lambda_s \geq 0.5\lambda_{pm}$, the matrix A_p is asymptotically stable, where $\lambda_{pm} = \min \mathrm{Re}\left(\lambda_{pi}\right)$ and λ_{pi} is the eigenvalue of $(L + B_L)$. Let $\Delta_p = \left[\Delta_{pi}\right] \in \mathbb{R}^{3N\times1}$ and $\Delta_a = \left[\Delta_{ai}\right] \in \mathbb{R}^{3N\times1}$. One can have that

$$\begin{aligned} \left\|e_p\right\|_\infty &\leq \mu_{p0}\lambda_0 + \pi_{p0} + \mu_p\left\|\Delta_p\right\|_\infty, \\ \left\|e_a\right\|_\infty &\leq \pi_{a0} + \mu_a\left\|\Delta_a\right\|_\infty, \end{aligned} \tag{5.21}$$

where

$$\begin{aligned} \pi_{p0} &= \max_j \sup_{t\geq0}\left|c_{6N,j}^T e^{A_p t} e_p(0)\right|, \\ \pi_{a0} &= \max_j \sup_{t\geq0}\left|c_{6N,j}^T e^{A_a t} e_a(0)\right|, \\ \lambda_0 &= \max_j \sup_{t\geq0}\left|\ddot{p}_{0,j}^r(t)\right|, \\ \mu_p &= \left\|\left(sI_{6N} - A_p\right)^{-1} B_{k2}\left(I_{3N} - F_p(s)\right)\right\|_1, \end{aligned}$$

$$\mu_{p0} = \left\| \left(sI_{6N} - A_p \right)^{-1} B_{k2} \right\|_1,$$

$$\mu_a = \left\| \left(sI_{6N} - A_a \right)^{-1} B_{k2} \left(I_{3N} - F_a(s) \right) \right\|_1,$$

$$\mu_a = \left\| \left(sI_{6N} - A_a \right)^{-1} B_{k2} \left(I_{3N} - F_a(s) \right) \right\|_1,$$

$F_p(s) = diag\{F_{pi}(s)\}$ and $F_a(s) = diag\{F_{ai}(s)\}$. According to [62, 101], there exist positive constants f_p^* and f_a^* such that if $f_{pm} \geq f_p^*$ and $f_{am} \geq f_a^*$, then one can have $\mu_p \leq \lambda_p / f_{pm}$, $\mu_a \leq \lambda_a / f_{am}$, and $\mu_{p0} \leq \lambda_{p0}$, where $f_{pm} = \min_{ij}\{f_{pij}\}$, $f_{am} = \min_{ij}\{f_{aij}\}(j = 1, 2, 3)$, and λ_p, λ_a, and λ_{p0} are positive constants.

Theorem 5.1

Consider the nonlinear dynamics of each tail-sitter given in (5.1) and the formation control protocol as shown in (5.12), (5.13), (5.17), and (5.18). If the directed graph G has a spanning tree, the root can receive the information from the virtual leader; then, for the initial bounded conditions $e_p(0)$ and $e_a(0)$ and given positive constants ε_p and ε_a, there exist positive constants f_p^*, f_a^*, and T^*, such that if $f_{pij} \geq f_p^*(j = 1, 2, 3)$ and $f_{aij} \geq f_a^*$, then the tracking errors involved in the global closed-loop control system are bounded and satisfy that $\max_j |e_{kpi,j}(t)| \leq \varepsilon_p$ and $\max_j |e_{kai,j}(t)| \leq \varepsilon_a, \forall t \geq T^*$.

Proof 5.1 From (5.10), one can have that

$$\begin{aligned} \left\| \Delta_{pi} \right\|_\infty &\leq \left\| B_{pi}^\Delta \right\|_1 \left\| R_i F_{ti} \right\|_\infty + \left\| B_{pi}^\Delta R_i \left(F_{wi} + F_{fi} \right) + m_i^{-1} R_i F_{di} \right\|_\infty \\ &\leq \chi_{\Delta pi1} \left\| u_{pi} \right\|_\infty + \chi_{\Delta pi2}. \end{aligned} \quad (5.22)$$

where $\chi_{\Delta pi1}$ and $\chi_{\Delta pi2}$ are positive constants. By substituting (5.12) and (5.13) into (5.11), one can obtain that

$$\left\| u_{pi} \right\|_\infty \leq \chi_{upi1} \left\| e_{kpi} \right\|_\infty + \chi_{upi2} \left\| \Delta_{pi} \right\|_\infty + \chi_{upi3}, \quad (5.23)$$

where χ_{upi1}, χ_{upi2}, and χ_{upi3} are positive constants. Combining (5.23) and (5.22) yields that

$$\left\| \Delta_{pi} \right\|_\infty \leq \chi_{\Delta ui1} \left\| e_{kpi} \right\|_\infty + \chi_{\Delta ui2}, \quad (5.24)$$

where $\chi_{\Delta ui1}$ and $\chi_{\Delta ui2}$ are positive constants. Let $\chi_{\Delta u1} = \max_i \chi_{\Delta ui1}$ and $\chi_{\Delta u2} = \max_i \chi_{\Delta ui2}$. From (5.24), one has that

$$\left\| \Delta_p \right\|_\infty \leq \chi_{\Delta u1} \left\| e_p \right\|_\infty + \chi_{\Delta u2}. \quad (5.25)$$

If $f_{pm} \geq f_p^*$, such that $f_{pm} \geq \chi_{\Delta u1}\lambda_p$, one can yield the following inequalities:

$$\begin{aligned} \left\|\Delta_p\right\|_\infty &\leq \frac{\chi_{\Delta u1}\left(\pi_{p0}+\lambda_{p0}\lambda_0\right)+\chi_{\Delta u2}}{1-\chi_{\Delta u1}\lambda_p f_{pm}^{-1}}, \\ \left\|e_p\right\|_\infty &\leq \frac{\pi_{p0}+\lambda_{p0}\lambda_0+\lambda_p f_{pm}^{-1}\chi_{\Delta u2}}{1-\chi_{\Delta u1}\lambda_p f_{pm}^{-1}}. \end{aligned} \tag{5.26}$$

Since π_{p0} is bounded and the matrix A_p is asymptotically stable, and the second derivative $\ddot{p}_0^r$ of the virtual leader trajectory is bounded and can converge to 0, there exist positive constants $\varepsilon_{\Delta p}$ and ε_{ep} such that

$$\begin{aligned} \left\|\Delta_p\right\|_\infty &\leq \varepsilon_{\Delta p}, \\ \left\|e_p\right\|_\infty &\leq \varepsilon_{ep}. \end{aligned} \tag{5.27}$$

From (5.21) and (5.27), one can obtain that

$$\max_k e_p(t) \leq \max_k \left|c_{6N,k}^T e^{A_p T} e_p(0)\right| + f_{pm}^{-1}\lambda_p \varepsilon_{\Delta p} + \lambda_{p0}\lambda_0. \tag{5.28}$$

From (5.28), one can see that for the initial bounded conditions $e_p(0)$ and $e_a(0)$ and given positive constants ε_p and ε_a, there exist positive constants f_p^* and T^*, such that if $f_{pij} \geq f_p^*\left(j=1,\ 2,\ 3\right)$, then the tracking errors involved in the global closed-loop control system are bounded and satisfy that $\max_j \left|e_{kpi,j}(t)\right| \leq \varepsilon_p$, $\forall t \geq T^*$.

Then, the robust properties of the attitude control system are proven. From (5.10), one can obtain that

$$\left\|\Delta_{ai}\right\|_\infty \leq \chi_{\Delta ai1}\left\|e_{kai}\right\|_\infty^2 + \chi_{\Delta ai2}\left\|e_{kai}\right\|_\infty + \chi_{\Delta ai3}\left\|u_{ai}\right\|_\infty + \chi_{\Delta ai4}, \tag{5.29}$$

where $\chi_{\Delta ai1}$, $\chi_{\Delta ai2}$, $\chi_{\Delta ai3}$, and $\chi_{\Delta ai4}$ are positive constants. From (5.16), (5.17), and (5.18), one can have that

$$\left\|u_{ai}\right\|_\infty \leq \chi_{uai1}\left\|e_{kai}\right\|_\infty + \chi_{uai2}\left\|\Delta_{ai}\right\|_\infty + \chi_{uai3}. \tag{5.30}$$

Combining (5.29) and (5.30) yields that

$$\left\|\Delta_{ai}\right\|_\infty \leq \chi_{\Delta ei1}\left\|e_{kai}\right\|_\infty^2 + \chi_{\Delta ei2}\left\|e_{kai}\right\|_\infty + \chi_{\Delta ei3}, \tag{5.31}$$

where $\chi_{\Delta ei1}$, $\chi_{\Delta ei2}$, and $\chi_{\Delta ei3}$ are positive constants. Let $\chi_{\Delta e1} = \max_i \chi_{\Delta ei1}$, $\chi_{\Delta e2} = \max_i \chi_{\Delta ei2}$, and $\chi_{\Delta e3} = \max_i \chi_{\Delta ei3}$. Then, from (5.31), one has that

$$\|\Delta_a\|_\infty \le \chi_{\Delta e1}\|e_a\|_\infty^2 + \chi_{\Delta e2}\|e_a\|_\infty + \chi_{\Delta e3}. \tag{5.32}$$

If the following inequality holds

$$\chi_{\Delta e1}\|e_a\|_\infty + \chi_{\Delta e2} \le 1\Big/\left(\sqrt{\mu_a} + \mu_a\right), \tag{5.33}$$

one can have the following equation by substituting (5.21) into (5.32) as:

$$\|\Delta_a\|_\infty \le \pi_{a0}\Big/\sqrt{\mu_a} + \left(1+\sqrt{\mu_a}\right)\chi_{\Delta e3}. \tag{5.34}$$

From (5.21) and (5.34), it follows that

$$\|e_a\|_\infty \le \pi_{a0}\left(1+\sqrt{\mu_a}\right) + \mu_a\chi_{\Delta e3}\left(1+\sqrt{\mu_a}\right). \tag{5.35}$$

For (5.33), one can obtain the attractive region of $e_a(t)$ as

$$\left\{e_a(t) : \|e_a\|_\infty \le \chi_{\Delta e1}^{-1}\left(\mu_a + \sqrt{\mu_a}\right)^{-1} - \chi_{\Delta e1}^{-1}\chi_{\Delta e2}\right\}. \tag{5.36}$$

From (5.34) and (5.21), one can have that

$$\max_k e_a(t) \le \max_k \left|c_{6N,k}^T e^{A_a T} e_a(0)\right| + \left(\pi_{a0} + \chi_{ea}\right)\sqrt{f_{am}^{-1}\lambda_a},$$

where $\chi_{ea} \ge \left(1+\sqrt{\mu_a}\right)\sqrt{\mu_a}\chi_{\Delta e3}$. From the above analysis, one can observe that there exist positive constants T^* and f_a^*, such that if $f_{aij} \ge f_a^*$ $(j = 1,2,3)$, the attitude tracking error of the global system satisfies that $\max_j \left|e_{kai,j}(t)\right| \le \varepsilon_a, \forall t \ge T^*$.

Remark 5.4

It should be pointed out that the theoretical values of the robust filter parameters f_{pij} and f_{aij} $(j = 1,\ 2,\ 3)$ determined by Theorem 5.1 may be conservative; that is, the actual values of f_{pij} and f_{aij} $(j = 1,\ 2,\ 3)$ may be much smaller than their theoretical values. Since the tracking performances can be improved by selecting f_{pij} and f_{aij} $(j = 1,\ 2,\ 3)$ with larger values, the robust filter parameters can be tuned online unidirectionally following this procedure. The first step is to set f_{pij} and f_{aij} $(j = 1,\ 2,\ 3)$ with a certain initial positive value. The second step is to increase the value of f_{pij} and f_{aij} $(j = 1,\ 2,\ 3)$ until the achieved attitude tracking performances are satisfied.

5.5 Simulation Results

In this section, the time-varying formation flight tasks are carried out to check the tracking performance of the proposed robust control method. The nominal model parameters of the tail-sitters are selected according to [84]. The desired trajectory of the virtual leader is given as: $p_{y0}^r = 0$, and

$$p_{x0}^r = \begin{cases} 0, \ t \leq 10, \\ t^2 - 20t + 100, \ 10 < t \leq 20, \\ 20t - 300, \ t > 20, \end{cases}$$

$$p_{z0}^r = \begin{cases} t, \ t \leq 10, \\ 40 - 30e^{-0.125(t-10)^2}, \ t > 10. \end{cases}$$

At the beginning of the formation mission, the four tail-sitters take off vertically in a diamond. Then, the formation pattern is converted from the diamond to one shape in the flight mode transition from vertical flight to level flight, which can save switching time and reduce flight resistance. The interaction topology of the tail-sitter formation system is modeled by a directed graph with the node set $V = \{v_1, v_2, v_3, v_4\}$, the edge set $E = \{(v_4, v_3), (v_3, v_2), (v_2, v_1)\}$, and the weighted matrix $W = [w_{ij}]$. If $(v_i, v_j) \in E$, then $w_{ij} = 1$ and $w_{ij} = 0$, otherwise. In the simulation tests, tail-sitter 1 can directly obtain the position and attitude information from the virtual leader; thereby, $k_{c1} = 1$, $k_{c2} = 0$, $k_{c3} = 0$, and $k_{c4} = 0$. Each real vehicle parameter is assumed to be 23% larger than its nominal value, respectively. The periodic and non-vanished external disturbances acting on tail-sitter i are given by

$$F_{di} = \begin{bmatrix} 7\sin(2t) + 11\cos(t) \\ 10\sin(3t) + 18\cos(t) \\ 9\sin(t) + 20\cos(2t) \end{bmatrix},$$

$$\tau_{di} = \begin{bmatrix} 7.5\sin(3t) + 2.4\cos(2t) \\ 7.5\sin(t) + 7\cos(3t) \\ 10\sin(2t) + 8\cos(3t) \end{bmatrix}.$$

In the flight task, the initial conditions of tail-sitters are given as $p_{I1}(0)=\begin{bmatrix}0 & 4 & 0\end{bmatrix}^T$, $p_{I2}(0)=\begin{bmatrix}-4 & 0 & 0\end{bmatrix}^T$, $p_{I3}(0)=\begin{bmatrix}0 & -4 & 0\end{bmatrix}^T$, $p_{I4}(0)=\begin{bmatrix}4 & 0 & 0\end{bmatrix}^T$, and $q_i(0)=\begin{bmatrix}\sqrt{2}/2 & 0 & \sqrt{2}/2 & 0\end{bmatrix}^T$. The nominal controller parameters are chosen as $K_a = diag\{149.5, 149.6, 149.6\}$, $K_{\dot{a}} = diag\{104.7, 149.6, 102.6\}$, $K_p = diag\{3.5, 3.5, 3.5\}$, and $K_{\dot{p}} = diag\{1.2, 1.2, 1.2\}$. The robust compensator parameters are determined as $f_{pi1} = 1000$, $f_{pi2} = 1000$, $f_{pi3} = 1000$, $f_{ai1} = 50$, $f_{ai2} = 50$, and $f_{ai3} = 50$, as shown in Remark 5.4. The time-varying formation flight of the four tail-sitters with the desired deviations can be determined by

$$\delta_1 = \begin{cases} [0 \quad 4 \quad 0]^T, \ t < 10, \\ \begin{bmatrix}0 & 6-2e^{-(t-10)} & 0\end{bmatrix}^T, \ t \geq 10, \end{cases}$$

$$\delta_2 = \begin{cases} [-4 \quad 0 \quad 0]^T, \ t < 10, \\ \begin{bmatrix}-4e^{-(t-10)} & 2-2e^{-(t-10)} & 0\end{bmatrix}^T, \ t \geq 10, \end{cases}$$

$$\delta_3 = \begin{cases} [0 \quad -4 \quad 0]^T, \ t < 10, \\ \begin{bmatrix}0 & 2e^{-(t-10)}-6 & 0\end{bmatrix}^T, \ t \geq 10, \end{cases}$$

$$\delta_4 = \begin{cases} [4 \quad 0 \quad 0]^T, \ t < 10, \\ \begin{bmatrix}4e^{-(t-10)} & 2e^{-(t-10)}-2 & 0\end{bmatrix}^T, \ t \geq 10. \end{cases}$$

Figure 5.3 illustrates the 3-D trajectories of the multiple tail-sitter system by the proposed robust control method. Figures 5.4 and 5.5 show the attitude and position response using the proposed robust method, respectively. The trajectory tracking error is shown in Figure 5.6. In contrast, the proposed robust controller is compared with the baseline nominal controller (see [55]) by ignoring the robust compensating inputs in (5.11) and (5.16). The attitude response is depicted in Figure 5.7. The trajectory tracking error by the baseline controller is shown in Figure 5.8. From these figures, one can see that the proposed robust controller can achieve the predefined aggressive continuous time-varying formation under the influence of the parametric perturbations, coupling, nonlinear dynamics, and external time-varying disturbances. The tracking performance of the proposed robust controller is improved compared to that by the baseline controller.

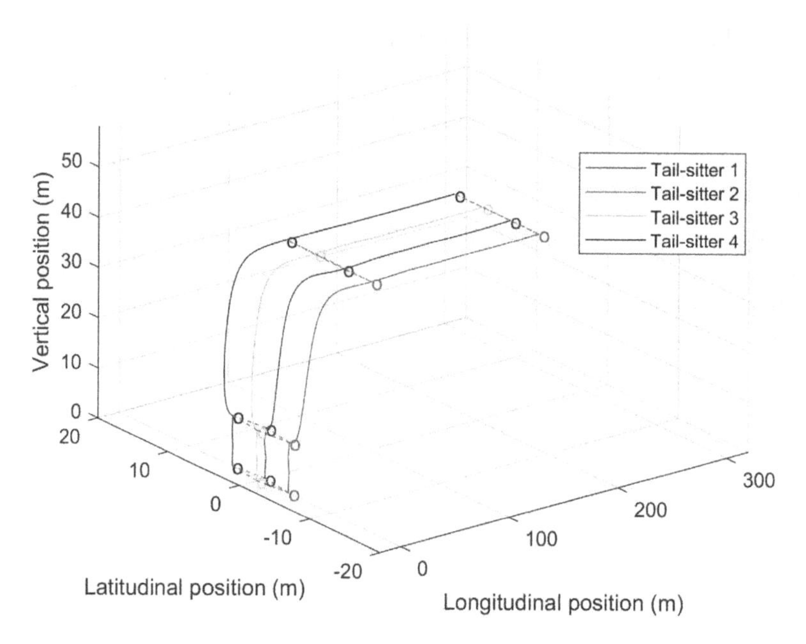

FIGURE 5.3
3-D trajectory tracking by the proposed robust controller.

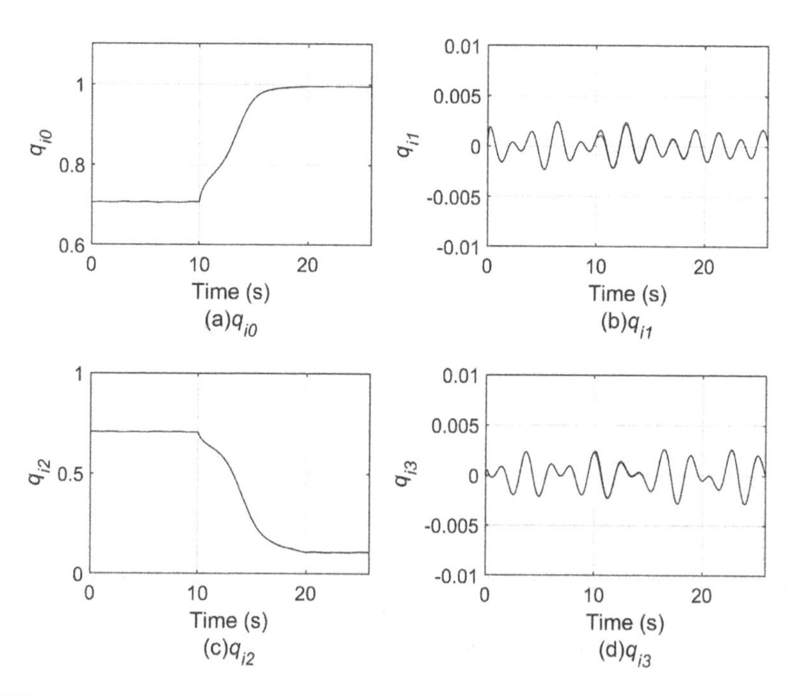

FIGURE 5.4
Attitude response of the proposed robust controller.

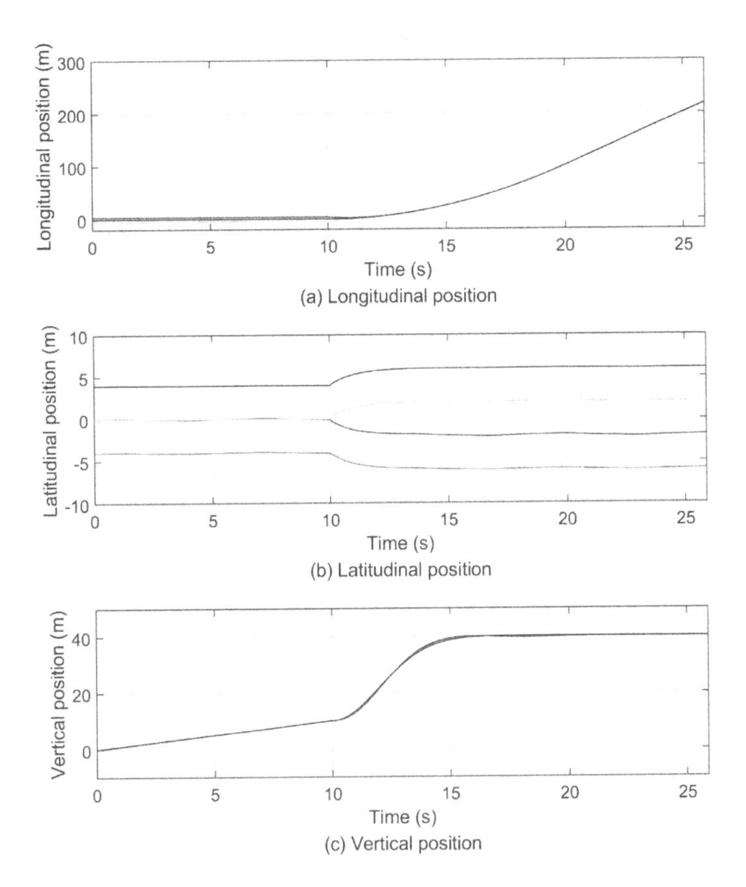

FIGURE 5.5
Position response of the proposed controller.

5.6 Conclusion

This chapter addresses the time-varying formation protocol design problem for a group of tail-sitters subject to underactuated, highly nonlinear and strongly coupled dynamics, and disturbances in aggressive flight mode transitions. A distributed robust formation controller is proposed based on the robust compensation theory. For each tail-sitter, the proposed controller results in a trajectory tracking controller and an attitude controller to govern the translational and rotational motions, respectively. The tracking errors of the proposed global closed-loop system can be guaranteed to converge to a given neighborhood of the origin in a finite time. Numerical simulation studies are provided to demonstrate the effectiveness of the proposed formation protocol scheme.

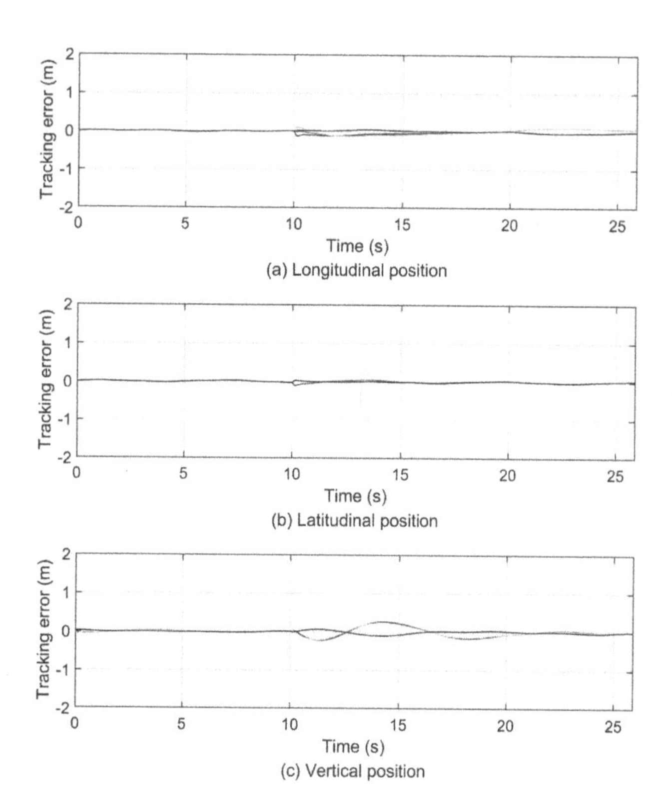

FIGURE 5.6
Trajectory tracking errors by the proposed robust controller.

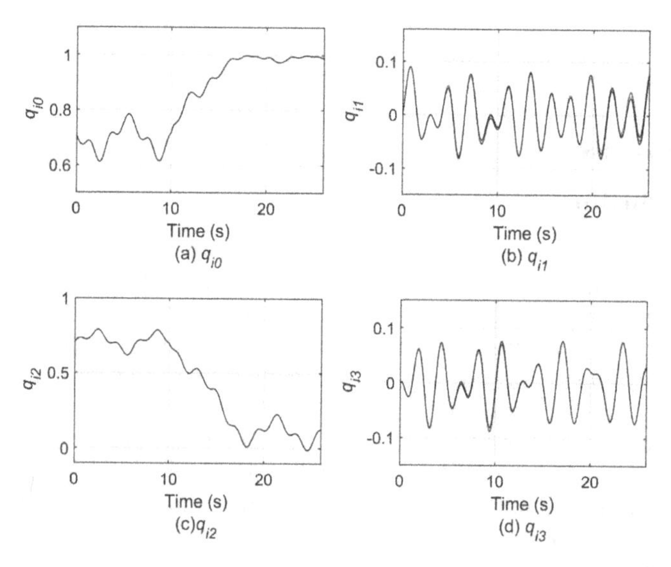

FIGURE 5.7
Attitude response of the baseline controller.

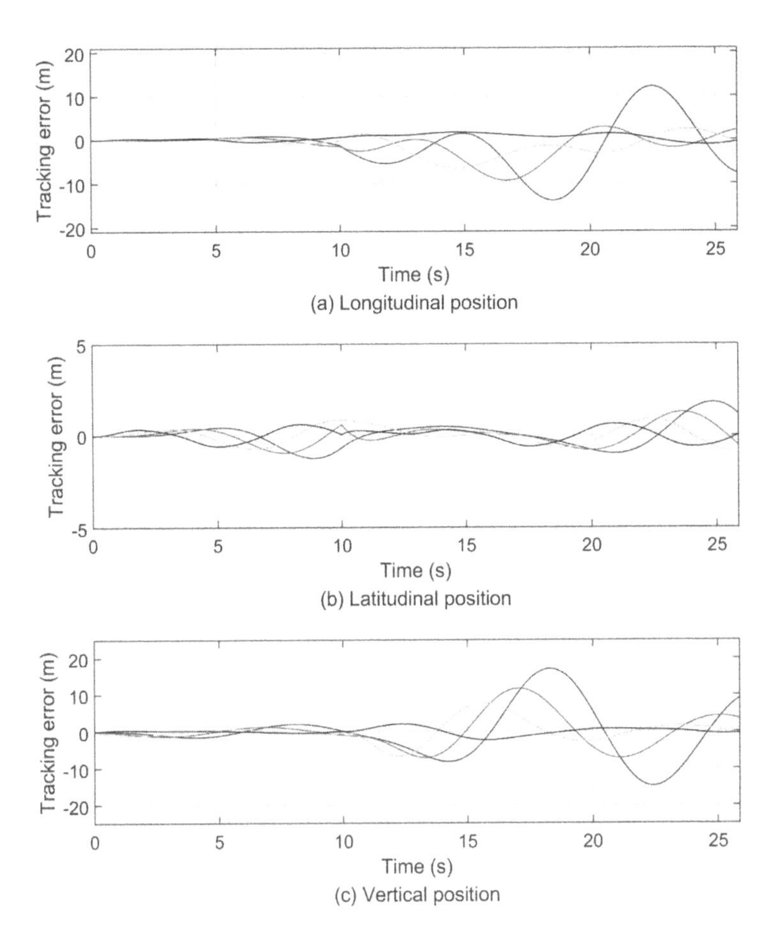

FIGURE 5.8
Trajectory tracking errors by the baseline controller.

6

Robust Fault-Tolerant Formation Control for Tail-Sitters in Aggressive Flight Mode Transitions

This chapter investigates the fault-tolerant time-varying formation control problem for a team of tail-sitters in the presence of unknown actuator faults and uncertainties. A robust distributed and continuous fault-tolerant formation control method is proposed to achieve aggressive time-varying formation flight in flight mode transitions. For each tail-sitter, the designed controller is divided into an outer position controller and an inner attitude controller to govern the translational and rotational motions, respectively. It is proven that the formation system stability can be guaranteed under the effects of actuator faults and uncertainties. Simulation results of the formation flight are given to show the effectiveness of the proposed control method.

6.1 Introduction

The formation control of unmanned aerial vehicles (UAVs) has gained considerable interest from various potential application fields, such as border surveillance, forest fire detection, and power line inspection (see [102–105] to mention a few). Cooperation among multiple UAVs has potential advantages of greater flexibility and higher performance over a single-body UAV. However, faults may occur frequently in the group of UAVs, and any failure may easily damage the formation system or even cause catastrophic accidents. Therefore, the safety and reliability of multi-vehicle systems are important and the development of a fault-tolerant formation control scheme is a necessity [106–111].

Recently, a special UAV concept, called tail-sitter UAVs, has been widely developed in the academic and industry as depicted in [83, 84, 86, 88, 112, 113]. Aerial vehicles have the abilities of high-speed cruising like a fixed-wing aircraft and vertical take-off and landing like a rotary-wing aircraft. They can perform in three flight modes: vertical flight, transition flight, and horizontal flight. Therefore, tail-sitter UAVs can accomplish a wider range of missions, compared to conventional fixed-wing aircrafts or rotary-wing aircrafts.

DOI: 10.1201/9781003242147-6

Multiple tail-sitters may operate following three formation flying modes ranging from vertical formation, transition formation, and cruising formation. For vertical formation, the take-off formation pattern may be adaptive to their surroundings, like the conventional quadrotor formation. In addition, a specific formation pattern can reduce air resistance in the cruising formation mode, like the traditional fixed-wing vehicle formation. In the transition formation mode, the formation pattern is required to change, and the team of tail-sitters needs to quickly enter the forward formation mode. However, the formation control protocol design is challenging for tail-sitters, especially in the transition formation mode. First, the tail-sitter formation pattern is required to perform the continuous flight mode transition, which is an aggressive flight process. Each tail-sitter is an under actuated complex system, involving parametric uncertainties, nonlinear and coupled dynamics, and external disturbances. Second, the tail-sitter flight controller is sensitive to multiple actuator faults in the flight mode transitions. Moreover, in the aggressive flight mode transitions, it is difficult to define the coordinate systems to describe the dynamics of the tail-sitters and design a continuous control law for different flight phases.

At present, many aforementioned approaches mainly focused on conventional UAV formation control with actuator faults. In [114], a consensus-based formation control method was proposed for multiple quadrotors subject to actuator faults. In [115], a robust adaptive control method was constructed for a group of typical quadrotors to achieve the desired time-invariant formation configuration under the effects of actuator faults. In [116], a fault-tolerant control strategy based on a leader-follower structure was developed to achieve a fixed formation pattern for a group of fixed-wing vehicles subject to actuator faults. In [117], a distributed sliding-mode controller was designed for multiple fixed-wing UAVs to deal with actuator faults. In [118], a fault-tolerant formation control design strategy was developed to overcome actuator faults in formation flight for a team of typical fixed-wing vehicles. In [119], a cooperative control strategy was developed for multiple 3-DOF helicopters with actuator faults to achieve time-invariant formation patterns. Although several fault-tolerant control approaches have been presented for conventional UAV formation flying, the time-varying formation control problem for the tail-sitters under flight mode transitions subject to multiple actuator faults and uncertainties is still unsolved.

Motivated by these considerations, this chapter proposes a robust fault-tolerant time-varying formation control method for the group of tail-sitters subject to multiple actuator faults and uncertainties in the flight mode transitions. The actuator faults of each tail-sitter are considered as a constant loss of effectiveness in the moments and forces generated by the actuators, and multiple simultaneous actuator faults of the tail-sitter formation system are studied. For each tail-sitter, the designed controller is divided into an outer position controller and an inner attitude controller to govern the translational and rotational motions, respectively. The new contributions of the proposed control approach in this chapter are summarized as follows.

First, a continuous time-varying formation pattern can be achieved in the flight mode transitions without requirements of switching controller structures or controller parameters. Second, each tail-sitter model considered in this chapter involves seriously nonlinear dynamics and high uncertainties involving parameter uncertainties and external disturbances. Third, the proposed distributed robust control protocol does not require identifying actuator faults online, and the effects of multiple actuator faults, parametric uncertainties, nonlinear and coupled dynamics, and external disturbances can be restrained simultaneously. The robustness properties can be guaranteed for the global closed-loop control system.

6.2 Problem Formulation

6.2.1 Aircraft Body

A schematic of the tail-sitter UAV is depicted in Figure 6.1. The vehicle is equipped with coaxial counter-rotating propellers to provide a dominant thrust or lift. Two wings are attached to the fuselage to generate enough lift for forward flight. Moreover, four small rotors mounted on the airframe tail and two ailerons situated at the trailing edge of the two wings simultaneously take charge of the attitude control in the mode transition flights.

Each tail-sitter can achieve the following three flight modes, as depicted in Figure 6.2. For vertical flight, the flight controller is similar to that of a

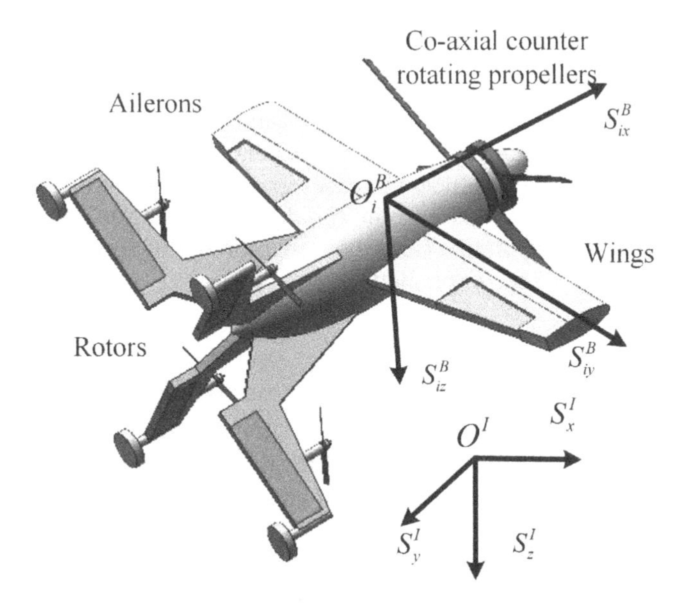

FIGURE 6.1
Schematic of a tail-sitter aircraft.

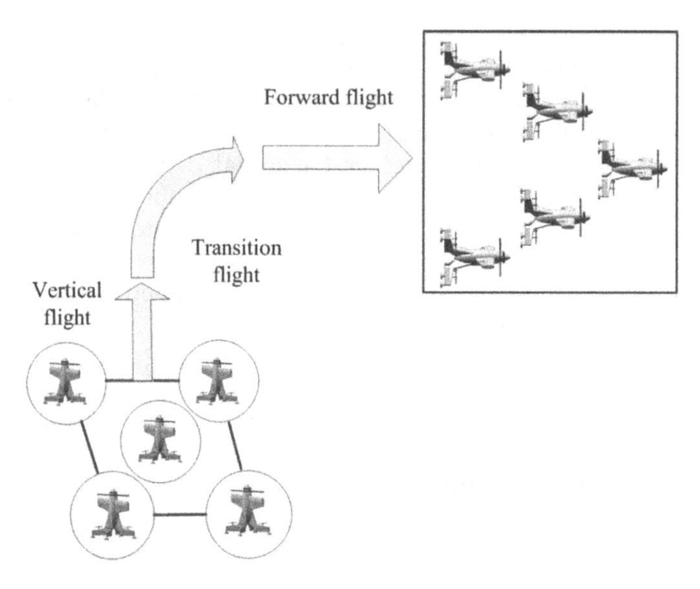

FIGURE 6.2
Flight path of tail-sitters.

quadrotor, i.e., the main lift force is generated by the coaxial counter-rotating propellers and the four rotors, while the attitude control is achieved by the differential force of the two pairs of rotors. In the forward flight, the dominant thrust is generated by the coaxial counter-rotating propellers and the lift is provided by the two wings. The yaw angle is controlled by the four rotors, while the pitch and roll angles are governed by the two ailerons. For the transition flight, in the initial phase of transition, the main lift is generated by the coaxial counter-rotating propellers, while the lift generated by the wings is not dominant. Then, the vertical component of the thrust provided by the coaxial counter-rotating propellers starts decreasing. As the forward flight speed of the vehicle increases gradually, the aerodynamic forces generated by the two wings become the dominant lift and can support the entire vehicle's weight. In this case, the transition flight from vertical flight to forward flight can be accomplished.

6.2.2 Dynamic Motion Equations

In this chapter, two frames are defined: the inertial frame $E^I = \{O^I, E_x^I, E_y^I, E_z^I\}$ attached to the earth to analyze the tail-sitter motion and the body-fixed frame $E_i^B = \{O_i^B, E_{ix}^B, E_{iy}^B, E_{iz}^B\}$ attached to the tail-sitter body.

From [84], the dynamical model of tail-sitter i can be given as

$$\begin{aligned} m_i\ddot{p}_i^I &= R_iF_{it} - m_igc_{3,3}, \\ J_i\dot{\omega}_i + \omega_i^{\times}J_i\omega_i &= M_{it}, \end{aligned} \tag{6.1}$$

where $p_i^I = \begin{bmatrix} p_{ix}^I & p_{iy}^I & p_{iz}^I \end{bmatrix}^T$ is the position vector with p_{ix}^I representing the longitudinal position, p_{iy}^I being the lateral position, and p_{iz}^I being the vertical position; m_i is the mass of the tail-sitter; F_{it} and M_{it} are the total force and the total moment acting on the tail-sitter, respectively, R_i is the orientation matrix from E_i^B to E^I, $J_i = diag\{J_{ix}, J_{iy}, J_{iz}\}$ is the inertia matrix, $c_{n,j}$ indicates an $n \times 1$ vector with one on the j-th row and zeros elsewhere, g represents the gravitation constant, $\omega_i = \begin{bmatrix} \omega_{ix} & \omega_{iy} & \omega_{iz} \end{bmatrix}^T$ is the angular velocity in E_i^B, and the superscript $\times$ represents the cross product. Because the Euler representation suffers from the singularity problem, the unit quaternion representation is applied to describe the rotational dynamics in the flight mode transitions. We define $q_i = \begin{bmatrix} q_{i0} & \bar{q}_i \end{bmatrix}^T$, where $\bar{q}_i = \begin{bmatrix} q_{i1} & q_{i2} & q_{i3} \end{bmatrix}^T$ and q_{i0} are the vector part and the scalar part of the quaternion, respectively, and satisfy $\bar{q}_i^T \bar{q}_i + q_{i0}^2 = 1$.

Therefore, the orientation matrix R_i by the unit quaternion is obtained by

$$R_i = \begin{bmatrix} 1 - 2q_{i1}^2 - 2q_{i3}^2 & 2q_{i1}q_{i2} - 2q_{i0}q_{i3} & 2q_{i1}q_{i3} + 2q_{i0}q_{i2} \\ 2q_{i1}q_{i2} + 2q_{i0}q_{i3} & 1 - 2q_{i1}^2 - 2q_{i3}^2 & 2q_{i2}q_{i3} - 2q_{i0}q_{i1} \\ 2q_{i1}q_{i3} - 2q_{i0}q_{i2} & 2q_{i2}q_{i3} + 2q_{i0}q_{i1} & 1 - 2q_{i1}^2 - 2q_{i2}^2 \end{bmatrix}.$$

The relationship between ω_i and q_i can be given as

$$\dot{\bar{q}}_i = 0.5\bar{q}_i^\times \omega_i + 0.5q_{i0}\omega_i,$$

$$\dot{q}_{i0} = -0.5\omega_i^T \bar{q}_i.$$

In practical flight, the total force F_{it} can be calculated as

$$F_{it} = F_{ir} + F_{io} + F_{iw} + F_{if} + F_{id}, \tag{6.2}$$

where F_{ir} is the total thrust by the four rotors, F_{io} represents the thrust generated by the coaxial counter-rotating propellers, F_{iw} is the aerodynamic force by the fixed wings, F_{if} is the aerodynamic force generated by the fuselage, and F_{id} indicates the external bounded disturbance force. The total moment M_{it} can be obtained as

$$M_{it} = M_{ir} + M_{ia} + M_{iw} + M_{ig} + M_{id}, \tag{6.3}$$

where M_{ir} is the torque generated by the small four rotors, M_{ia} represents the torque generated by the two ailerons, M_{iw} indicates the aerodynamic moment generated by the fixed wings, M_{ig} is the gyroscopic moment generated by the high-speed rotation of the rotors, and M_{id} is the external bounded disturbance torque. The thrusts F_{ir} and F_{io} can be calculated as follows:

$$F_{ir} = \left[k_{ir} \sum_{j=1}^{4} \omega_{ij}^2 \quad 0 \quad 0 \right]^T, \; F_{io} = \left[k_{io} \omega_{io}^2 \quad 0 \quad 0 \right]^T, \tag{6.4}$$

where k_{ir} and k_{io} denote two coefficients, ω_{ij} is the rotor rotational speed, and ω_{io} represents the rotational speed of the coaxial counter-rotating propellers. The thrusts F_{ir} and F_{io} satisfy $F_{io} = k_{it} F_{ir}$, where k_{it} is a positive constant. F_{iw} and F_{if} can be described as

$$F_{iw} = \begin{bmatrix} (L_{i1} + L_{i2}) \sin \alpha_i - (D_{i1} + D_{i2}) \cos \alpha_i \\ 0 \\ -(L_{i1} + L_{i2}) \cos \alpha_i - (D_{i1} + D_{i2}) \sin \alpha_i \end{bmatrix}, \; F_{if} = \begin{bmatrix} L_{if} \sin \alpha_i - D_{if} \cos \alpha_i \\ 0 \\ -L_{if} \cos \alpha_i - D_{if} \sin \alpha_i \end{bmatrix}, \tag{6.5}$$

where L_{i1} and L_{i2} are the lift forces produced by the wings, D_{i1} and D_{i2} are the drag forces generated by the wings, and L_{if} and D_{if} are the lift and drag forces produced by the fuselage, respectively. L_{i1}, L_{i2}, D_{i1}, D_{i2}, L_{if}, and D_{if} can be given as follows:

$$L_{ij} = 0.5 C_{i,Lj} S_{iw} \rho \left(\left(v_{ix}^B \right)^2 + \left(v_{iz}^B \right)^2 \right),$$

$$D_{ij} = 0.5 C_{i,Dj} S_{iw} \rho \left(\left(v_{ix}^B \right)^2 + \left(v_{iz}^B \right)^2 \right),$$

$$L_{if} = 0.5 C_{i,lf} S_{if} \rho \left(\left(v_{ix}^B \right)^2 + \left(v_{iz}^B \right)^2 \right),$$

$$D_{if} = 0.5 C_{i,df} S_{if} \rho \left(\left(v_{ix}^B \right)^2 + \left(v_{iz}^B \right)^2 \right), \; j = 1, \; 2,$$

where $C_{i,Lj}$, $C_{i,Dj}$, $C_{i,lf}$, and $C_{i,df}$ are the lift or drag coefficients, S_{iw} is the main wing area, ρ is the density of air, and S_{if} is the blade area. $C_{i,Lj}$, $C_{i,Dj}$, $C_{i,lf}$, and $C_{i,df}$ can be obtained as

$$C_{i,Lj} = C_{i,L0} + C_{i,L\alpha} \alpha_i + C_{i,L\delta} \delta_{ij},$$

$$C_{i,Dj} = C_{i,D0} + C_{i,Lj}^2 / A_{iD},$$

$$C_{i,lf} = C_{i,lf} \alpha_i,$$

$$C_{i,df} = C_{i,f0} + C_{i,f\alpha} \alpha_i, \; j = 1, \; 2,$$

where $C_{i,L0}$, $C_{i,L\alpha}$, $C_{i,L\delta}$, $C_{i,D0}$, $C_{i,lf}$, $C_{i,f0}$, and $C_{i,f\alpha}$ are the aerodynamic coefficients, δ_{ij} is the flap bias angle of the aileron, and A_{iD} is a positive constant. M_{ir} is calculated according to

$$M_{ir} = \begin{bmatrix} M_{i,r1} \\ M_{i,r2} \\ M_{i,r3} \end{bmatrix} = \begin{bmatrix} \sum_{j=1}^{4} (-1)^{j+1} \left(a_{i1} + \sqrt{2} a_{i2} l_{i1}/2\right) \omega_{ij}^2 \\ l_{i1} \left[F_{i,r1} + F_{i,r2} - F_{i,r3} - F_{i,r4}\right]/2 \\ l_{i1} \left[-F_{i,r1} + F_{i,r2} + F_{i,r3} - F_{i,r4}\right]/2 \end{bmatrix}, \tag{6.6}$$

where a_{i1} and a_{i2} are positive constants and l_{i1} represents the distance between two rotors. M_{iw} can be given as follows:

$$M_{iw} = \begin{bmatrix} l_{i2}(L_{i2} - L_{i1})\cos\alpha_i + l_{i2}(D_{i2} - D_{i1})\sin\alpha_i \\ l_{i3}(L_{i2} + L_{i1})\cos\alpha_i + l_{i3}(D_{i2} + D_{i1})\sin\alpha_i \\ l_{i2}(L_{i2} - L_{i1})\sin\alpha_i + l_{i2}(D_{i1} - D_{i2})\cos\alpha_i \end{bmatrix}, \tag{6.7}$$

where l_{i2} is the distance from the gravity center of the tail-sitter to the force acting position of the fixed wing and l_{i3} is the distance between the gravity center of the tail-sitter and the center position of the two wings. M_{ia} can be obtained by

$$M_{ia} = \begin{bmatrix} M_{i,a1} \\ M_{i,a2} \\ M_{i,a3} \end{bmatrix} = \begin{bmatrix} l_{i2} C_{i,L\delta} S_{iw} \rho \cos\alpha_i \left(\left(v_{ix}^B\right)^2 + \left(v_{iz}^B\right)^2 \right) (\delta_{i1} - \delta_{i2}) \\ l_{i3} C_{i,L\delta} S_{iw} \rho \cos\alpha_i \left(\left(v_{ix}^B\right)^2 + \left(v_{iz}^B\right)^2 \right) (\delta_{i1} + \delta_{i2}) \\ 0 \end{bmatrix}. \tag{6.8}$$

M_{ig} can be calculated as follows:

$$M_{ig} = \sum_{j=1}^{4} J_{ir} \omega_i^{\times} c_{3,3} (-1)^j \omega_{ij}, \tag{6.9}$$

where J_{ir} is the rotating inertia of each rotor. In the actual flight, the model parameters show deviation relative to theoretical values. All of the vehicle parameters are divided into two parts, for example, $J_i = J_i^N + J_i^\Delta$, where the superscript N indicates the nominal parameter and the superscript Δ indicates the parameter uncertainty. Then, the dynamical model in (6.1) can be rewritten as

$$\begin{aligned} m_i^N \ddot{p}_i^I &= R_i F_{it} - m_i^N g c_{3,3} + \Delta_{im}, \\ J_i^N \dot{\omega}_i &= -\omega_i^{\times} J_i^N \omega_i + M_{it} + \Delta_{iJ}, \end{aligned} \tag{6.10}$$

where the uncertainty terms involving the parameter uncertainties satisfy $\Delta_{im} = -m_i^\Delta g c_{3,3} - m_i^\Delta \ddot{p}_i^I$ and $\Delta_{iJ} = -J_i^\Delta \dot{\omega}_i - \omega_i^{\times} J_i^\Delta \omega_i$.

6.2.3 Actuators

The main control forces and torques of the tail-sitters are generated by four small rotors and two ailerons. Moreover, coaxial counter-rotating propellers are installed on the top of the airframe to generate the dominant lift force or the thrust force.

The reliability of the tail-sitter is improved. For instance, if the rotors are damaged or have performance degradation, other actuators can still govern the octorotor flying. In practical application, the actuators may suffer from faults (partial loss or complete loss of the control forces and torques). The control moment M_{ic} generated by the four rotors and the two ailerons can be calculated by $M_{ic} = M_{ir} + M_{ia}$. The force F_{ic} generated by the rotors and the coaxial counter-rotating propellers can be obtained by $F_{ic} = F_{ir} + F_{io}$. Therefore, M_{ic} and F_{ic} can be written as

$$M_{ic} = \Lambda_{i1}U_{i1} + \Lambda_{i2}U_{i2},$$

$$F_{ic} = \Lambda_{i3}U_{i1},$$

where $U_{i1} = \begin{bmatrix} \omega_{i1}^2 & \omega_{i2}^2 & \omega_{i3}^2 & \omega_{i4}^2 \end{bmatrix}^T$, $U_{i2} = \begin{bmatrix} \delta_{i1} & \delta_{i2} & 0 & 0 \end{bmatrix}^T$, and

$$\Lambda_{i1} = \begin{bmatrix} k_{i,\omega 1} & -k_{i,\omega 1} & k_{i,\omega 1} & -k_{i,\omega 1} \\ k_{i,\omega 2} & k_{i,\omega 2} & -k_{i,\omega 2} & -k_{i,\omega 2} \\ -k_{i,\omega 3} & k_{i,\omega 3} & k_{i,\omega 3} & -k_{i,\omega 3} \end{bmatrix},$$

$$\Lambda_{i2} = \begin{bmatrix} k_{i,a1} & -k_{i,a1} & 0 & 0 \\ k_{i,a2} & k_{i,a2} & 0 & 0 \\ 0 & 0 & 0 & 0 \end{bmatrix},$$

$$\Lambda_{i3} = \begin{bmatrix} k_{i,ro} & k_{i,ro} & k_{i,ro} & k_{i,ro} \\ 0 & 0 & 0 & 0 \\ 0 & 0 & 0 & 0 \end{bmatrix},$$

with $k_{i,a1} = l_{i2}C_{i,L\delta}S_{iw}\rho\cos\alpha_i\left(\left(v_{ix}^B\right)^2 + \left(v_{iz}^B\right)^2\right)$, $k_{i,\omega 1} = a_{i1} + \sqrt{2}a_{i2}l_{i1}/2$, $k_{i,a2} = l_{i3}C_{i,L\delta}S_{iw}\rho\cos\alpha_i\left(\left(v_{ix}^B\right)^2 + \left(v_{iz}^B\right)^2\right)$, $k_{i,\omega 2} = k_{ir}l_{i1}/2$, $k_{i,\omega 3} = k_{ir}l_{i1}/2$, and $k_{i,ro} = k_{ir}(1 + k_{it})$.

In this chapter, the actuator faults represent constant partial loss of effectiveness, which means that the faults are characterized by a decrease in the control moments and forces from their nominal values. Therefore, the actuator model can be written as

$$\Gamma_{i,r1} = diag\{\sigma_{i,r1}, \sigma_{i,r2}, \sigma_{i,r3}, \sigma_{i,r4}\},$$
$$\Gamma_{i,a1} = diag\{\sigma_{i,a1}, \sigma_{i,a2}, 0, 0\},\ \sigma_{i,rj}, \sigma_{i,ak} \in (0,1],\ j = 1,2,3,4,\ k = 1,2,$$

where $\sigma_{i,rj}$ and $\sigma_{i,ak}$ represent the loss of effectiveness fault gain of the control moments or forces after faults. For example, $\sigma_{i,r1} = 0.3$, which means that the control torque and thrust generated by Rotor 1 lose 30%. Then, the actuator model can be described as follows:

$$\begin{aligned} M_{ic} &= M_{i,c0} - \Lambda_{i1}\Gamma_{i,r1}U_{i1} - \Lambda_{i2}\Gamma_{i,a1}U_{i2}, \\ F_{ic} &= F_{i,c0} - \Lambda_{i3}\Gamma_{i,r1}U_{i1}, \end{aligned} \tag{6.11}$$

where $M_{i,c0} = \Lambda_{i1}U_{i1} + \Lambda_{i2}U_{i2}$ and $F_{i,c0} = \Lambda_{i3}U_{i1}$ are the control moment and force without faults, respectively.

6.2.4 Problem Statement

Let $B_{ip} = (m_i^N)^{-1} I_3$ and $B_{ia} = (J_i^N)^{-1}$. We define $u_{ip} = R_i F_{i.c0}$ as the virtual position control input and $u_{ia} = M_{i,c0}$ as the attitude control input. From (6.10), one can have the following system dynamics:

$$\begin{aligned} \ddot{p}_i^I &= B_{ip}u_{ip} + B_{ip}R_i(F_{iw} + F_{if}) - gc_{3,3} + \Delta_{ip}, \\ \dot{\omega}_i &= -B_{ia}\omega_i^{\times}J_i^N\omega_i + B_{ia}u_{ia} + B_{ia}(M_{iw} + M_{ig}) + \Delta_{ia}, \end{aligned} \tag{6.12}$$

where Δ_{ip}, Δ_{ia} are the equivalent disturbances and have the forms:

$$\begin{aligned} \Delta_{ip} &= B_{ip}R_iF_{id} - B_{ip}R_i\Lambda_{i3}\Gamma_{i,r1}U_{i1} + B_{ip}\Delta_{im}, \\ \Delta_{ia} &= B_{ia}M_{id} - B_{ia}(\Lambda_{i3}\Gamma_{i,r1}U_{i1} + \Lambda_{i2}\Gamma_{i,a1}U_{i2}) + B_{ia}\Delta_{iJ}. \end{aligned} \tag{6.13}$$

The control objective in this chapter is to design a robust distributed fault-tolerant formation control protocol to achieve the aggressive time-varying formation flight under multiple actuator faults and uncertainties in the flight mode transitions. A virtual leader is introduced to provide the desired formation flight trajectory. Let $p_0^r = \begin{bmatrix} p_{x0}^r & p_{y0}^r & p_{z0}^r \end{bmatrix}^T \in \mathbb{R}^{3\times1}$ be the bounded reference trajectory of a virtual leader, where its second derivative $\ddot{p}_0^r$ is assumed to be bounded and can converge to 0 in a finite time. We define $h_{ij} = \begin{bmatrix} h_{xij} & h_{yij} & h_{zij} \end{bmatrix}^T \in \mathbb{R}^{3\times1}$ as the desired time-varying position deviation between tail-sitter i and tail-sitter j, which determines the time-varying formation pattern of the tail-sitter group. Let h_i represent the position deviation between tail-sitter i and the virtual leader, satisfying $h_{ij} = h_i - h_j$.

Remark 6.1

It can be observed that the tail-sitter dynamics includes serious nonlinearities, parametric perturbation, and airflow disturbances, as shown in (6.10) and (6.11), and is more prone to actuator fault effects. Therefore, it is challenging to construct a robust fault-tolerant formation controller for a group of tail-sitters. By ignoring Δ_{ip} and Δ_{ia}, the model in (6.12) represents the nominal model. The real vehicle model can be regarded as the nominal model added with Δ_{ip} and Δ_{ia}.

6.3 Robust Controller Design

In this section, a robust distributed fault-tolerant time-varying formation controller is proposed for a team of tail-sitters, which consists of an outer position controller and an inner attitude controller for each tail-sitter.

6.3.1 Outer Position Controller Design

For tail-sitter i, the position control input u_{ip} is constructed as including the following two parts:

$$u_{ip} = u_{ip}^{N} + u_{ip}^{R}. \tag{6.14}$$

where u_{ip}^{N} is the nominal control input and u_{ip}^{R} is the robust compensating input. The nominal controller is constructed to achieve the desired formation control for the nominal translational system, while the robust compensation controller is constructed to restrain the effects of the equivalent disturbance Δ_{ip}. The nominal controller is designed as

$$\begin{aligned} u_{ip}^{N} = & -\lambda_g \sum_{j \in N_i} w_{ij} B_{ip}^{-1} \left(K_p \left(p_i^I - p_j^I - h_{ij} \right) + K_v \left(\dot{p}_i^I - \dot{p}_j^I - \dot{h}_{ij} \right) \right) \\ & - \lambda_g B_{ip}^{-1} k_{ig} \left(K_p \left(p_i^I - \dot{h}_i - p_0^r \right) + K_v \left(\dot{p}_i^I - \dot{h}_i - \dot{p}_0^r \right) \right) \\ & - R_i \left(F_{iw} + F_{if} \right) + B_{ip}^{-1} g c_{3,3} + B_{ip}^{-1} \ddot{h}_i, \end{aligned} \tag{6.15}$$

where λ_g indicates a positive coupling gain, K_p and K_v are the nominal controller gain matrices, and k_{ig} is the connection weight between tail-sitter i and the virtual leader. $k_{ig} > 0$ means that tail-sitter i can access the information of the virtual leader; otherwise $k_{ig} = 0$.

Furthermore, u_{ip}^{R} is constructed to overcome the influence of the multiple actuator faults and uncertainties. In (6.13), one can see that the influence of the multiple actuator faults and uncertainties is considered in Δ_{ip}. Because Δ_{ip} involves multiple uncertain terms and thereby cannot be measured directly in practical applications, the robust filters $F_{ip}(s) = diag\{F_{i,p1}(s), F_{i,p2}(s), F_{i,p3}(s)\}$ are introduced here, where $F_{i,pj}(s) = f_{i,pj}^2 / (s + f_{i,pj})^2 \; (j = 1,\ 2,\ 3)$, s is the Laplace operator, and $f_{i,pj}$ is a positive robust filter parameter. Therefore, we design the robust compensating input u_{ip}^R as:

$$u_{ip}^R(s) = -B_{ip}^{-1} F_{ip}(s) \Delta_{ip}(s). \tag{6.16}$$

One can observe that robust filters have the following property. $F_{i,pj}(s)\;(j = 1,\ 2,\ 3)$ have wider frequency bandwidths by selecting larger robust compensator parameters $f_{i,pj}\;(j = 1,\ 2,\ 3)$. Then, the robust filter gains are closer to 1. In this case, the influence of the equivalent disturbance Δ_{ip} can be restrained. In (6.12), one can obtain that

$$\Delta_{ip} = \ddot{p}_i^I - B_{ip} u_{ip} - B_{ip} R_i \left(F_{iw} + F_{if} \right) + g c_{3,3}. \tag{6.17}$$

From (6.16) and (6.17), the robust compensator can be realized as follows:

$$\begin{aligned} \dot{\eta}_{i,p1} &= -f_{ip} \eta_{i,p1} - f_{ip}^2 p_i^I + B_{ip} u_{ip} + B_{ip} R_i \left(F_{iw} + F_{if} \right) - g c_{3,3}, \\ \dot{\eta}_{i,p2} &= -f_{ip} \eta_{i,p2} + 2 f_{ip} p_i^I + \eta_{i,p1}, \\ u_{ip}^R &= B_{ip}^{-1} f_{ip}^2 \left(\eta_{i,p2} - p_i^I \right), \end{aligned} \tag{6.18}$$

where $\eta_{i,p1}$ and $\eta_{i,p2}$ are the filter states and $f_{ip} = diag\left(f_{i,p1}, f_{i,p2}, f_{i,p3} \right)$.

6.3.2 Inner Attitude Controller Design

The attitude controller of each tail-sitter is designed using only the information of itself. Define the attitude tracking error e_{ia} as

$$e_{ia} = \begin{bmatrix} e_{iQ} & e_{i\omega} \end{bmatrix}^T = \begin{bmatrix} \tilde{Q}_i & \omega_i - \omega_i^r \end{bmatrix}^T, \tag{6.19}$$

where $\tilde{Q}$ is a nonlinear function and ω_i^r is the desired angular velocity. Denote the desired attitude reference as $q_i^r = [q_{i0}^r \;\; q_{i1}^r \;\; q_{i2}^r \;\; q_{i3}^r]^T$. The nonlinear function $\tilde{Q}$ can be given as follows:

$$\tilde{Q}(q_i, q_i^r) = 2\,\mathrm{sgn}\left(\sum_{j=0}^{3} q_{ij} q_{ij}^r \right) \begin{bmatrix} -q_{i0} q_{i1}^r + q_{i1} q_{i0}^r + q_{i2} q_{i3}^r - q_{i3} q_{i2}^r \\ -q_{i0} q_{i2}^r - q_{i1} q_{i3}^r + q_{i2} q_{i0}^r + q_{i3} q_{i1}^r \\ -q_{i0} q_{i3}^r + q_{i1} q_{i2}^r - q_{i2} q_{i1}^r + q_{i3} q_{i0}^r \end{bmatrix}.$$

The desired angular velocity ω_i^r can be obtained by $\omega_i^r = 0.5\begin{bmatrix} -\bar{q}_i^r & \left(q_{i0}^r I_3 - \left(\bar{q}_i^r\right)^{\times}\right)^T \end{bmatrix} \dot{q}_i^r$. From [120], the error dynamics of the attitude system can be given as $\dot{e}_{ia} = \begin{bmatrix} \omega_i - \omega_i^r & \dot{\omega}_i - \dot{\omega}_i^r \end{bmatrix}^T$.

Similarly to u_{ip}, the attitude control input u_{ia} can be designed with two parts: the nominal control input u_{ia}^N and the robust compensating input u_{ia}^R, as follows:

$$u_{ia} = u_{ia}^N + u_{ia}^R. \tag{6.20}$$

We design the nominal control input u_{ia}^N as follows:

$$u_{ia}^N = B_{ia}^{-1}\left(-K_q e_{iQ} - K_w e_{iw} + \dot{\omega}_i^r\right) + \omega_i^{\times} J_i^N \omega_i - M_{iw} - M_{ig}, \tag{6.21}$$

where K_q and K_w are diagonal positive gain matrices. One can construct the robust compensator u_{ia}^R as

$$u_{ia}^R(s) = -B_{ia}^{-1} F_{ia}(s) \Delta_{ia}(s), \tag{6.22}$$

where $F_{ia}(s) = diag\left\{F_{i,a1}(s), F_{i,a2}(s), F_{i,a3}(s)\right\}$ and $F_{i,aj}(s) = f_{i,aj}^2 / \left(s + f_{i,aj}\right)^2$ $(j = 1,2,3)$. The realization of the robust compensating input u_{ia}^R is similar to that of u_{ip}^R.

Remark 6.2

One can see that *the designed controller is continuous in the flight mode transitions, without any switching requirements on the coordinate systems, controller structures, or controller parameters. Moreover, the proposed fault-tolerant time-varying formation control strategy is distributed because the designed control law of each tail-sitter depends on the position and velocity information from its neighbors and itself.*

6.4 Robust Property Analysis

In this section, the robustness stability and tracking properties of our overall constructed control system can be guaranteed in the presence of actuator faults and multiple uncertainties using the small gain theorem.

We define the position tracking error as $e_i^p = p_i^I - h_i - p_0^r$. Let $e_{ip} = \begin{bmatrix} \left(e_i^p\right)^T & \left(\dot{e}_i^p\right)^T \end{bmatrix}^T$. According to (6.12), (6.15), (6.16), (6.21), and (6.22), one has that

$$\dot{e}_{ip} = A_{ip}e_{ip} - \lambda_g B_{k1}\sum_{j\in N_i} w_{ij}\left(K_p\left(e_i^p - e_j^p\right) + K_v\left(\dot{e}_i^p - \dot{e}_j^p\right)\right)$$
$$- \lambda_g k_{ig} B_{k1}\left(K_p e_i^p + K_v \dot{e}_i^p\right) + B_{k1}\left(B_{ip}u_{ip}^R + \Delta_{ip} + \ddot{p}_0^r\right), \quad (6.23)$$
$$\dot{e}_{ia} = A_{ia}e_{ia} + B_{k1}\left(B_{ia}u_{ia}^R + \Delta_{ia}\right),$$

where

$$A_{ip} = \begin{bmatrix} 0_{3\times3} & I_3 \\ 0_{3\times3} & 0_{3\times3} \end{bmatrix}, A_{ia} = \begin{bmatrix} 0_{3\times3} & I_3 \\ -K_q & -K_w \end{bmatrix}, B_{k1} = \begin{bmatrix} 0_{3\times3} \\ I_3 \end{bmatrix}.$$

Thus, the global closed-loop error system can be written as

$$\dot{e}_p = A_p e_p + B_{k2}\left(B_{ip}u_{ip}^R + \Delta_{ip} + \ddot{p}_0^r\right),$$
$$\dot{e}_a = A_a e_a + B_{k2}\left(B_{ia}u_{ia}^R + \Delta_{ia}\right), \quad (6.24)$$

where $e_p = \left[e_{ip}\right] \in \mathbb{R}^{6N\times1}$, $e_a = \left[e_{ia}\right] \in \mathbb{R}^{6N\times1}$, and

$$A_p = I_N \otimes A_{ip} - \lambda_g\left(L + B_L\right) \otimes B_k K_t,$$
$$A_q = I_N \otimes A_{ia},\ B_{k2} = I_N \otimes B_{k1},$$

with $B_L = diag\left\{k_{ig}\right\} \in \mathbb{R}^{N\times N}$ and $K_t = \left[K_p \quad K_v\right]$. One can observe that the matrix A_q is asymptotically stable by choosing proper gain matrices K_q and K_w. From [120], if the directed graph G involves a spanning tree and there exists a root that can receive the information from the virtual leader, the matrix A_p is asymptotically stable, if $\lambda_g \geq 0.5\lambda_{pm}$, where $\lambda_{pm} = \min \mathrm{Re}\left(\lambda_{ip}\right)$ and λ_{ip} is the eigenvalues of $\left(L + B_L\right)$.

From (6.24), one can obtain

$$\left\|e_p\right\|_\infty \leq \chi_{p0}\chi_0 + \left\|\pi_{p0}\right\|_\infty + \chi_p\left\|\Delta_p\right\|_\infty,$$
$$\left\|e_a\right\|_\infty \leq \left\|\pi_{a0}\right\|_\infty + \chi_a\left\|\Delta_a\right\|_\infty, \quad (6.25)$$

where $\Delta_p = \left[\Delta_{ip}\right] \in \mathbb{R}^{3N\times1}$ and $\Delta_a = \left[\Delta_{ia}\right] \in \mathbb{R}^{3N\times1}$. $\pi_{p0} = e^{A_p t}e_p(0)$, $\pi_{a0} = e^{A_a t}e_a(0)$, $\chi_p = \left\|\left(sI_{6N} - A_p\right)^{-1} B_{k2}\left(I_{3N} - F_p(s)\right)\right\|_1$, $\chi_{p0} = \left\|\left(sI_{6N} - A_p\right)^{-1} B_{k2}\right\|_1$, $\chi_a = \left\|\left(sI_{6N} - A_a\right)^{-1} B_{k2}\left(I_{3N} - F_a(s)\right)\right\|_1$, $\chi_0 = \max_j\left|\ddot{p}_{0,j}^r(t)\right|$, $F_p(s) = diag\left\{F_{ip}(s)\right\}$, and $F_a(s) = diag\left\{F_{ia}(s)\right\}$. From [49], there exist positive constants f_p^* and f_a^* such that if $f_{pm} \geq f_p^*$ and $f_{am} \geq f_a^*$, then one can have $\chi_p \leq \lambda_p / f_{pm}$, $\chi_a \leq \lambda_a / f_{am}$ and

$\chi_{p0} \leq \lambda_{p0}$, where $f_{pm} = \min_j \{f_{i,pj}\}$, $f_{am} = \min_j \{f_{i,aj}\}(j = 1, 2, 3)$, and λ_p, λ_a, and λ_{p0} are positive constants.

Theorem 6.1

Consider the tail-sitter formation system described by (6.1) and the robust formation control method in Section 6.3. If the directed graph G has a spanning tree, the root can receive the information from the virtual leader; then, for the initial bounded conditions, $e_p(0)$ and $e_a(0)$ and given positive constants ε_p and ε_a, there exist positive constants f_p^*, f_a^*, T_p^*, and T_a^*, such that if $f_{i,pj} \geq f_p^*(j = 1, 2, 3)$ and $f_{i,aj} \geq f_a^*$, then the time-varying formation flight of multiple tail-sitters can be achieved and the tracking errors are bounded and satisfy that $\max_j |e_{i,pj}(t)| \leq \varepsilon_p, \forall t \geq T_p^*$ and $\max_j |e_{i,aj}(t)| \leq \varepsilon_a, \forall t \geq T_a^*$.

Proof 6.1 From (6.13), one can have that

$$\begin{aligned} \|\Delta_{ip}\|_\infty &\leq \|B_{ip}\|_1 \|R_i \Lambda_{i3} \Gamma_{i,r1} U_{i1}\|_\infty + \|B_{ip}\|_1 \|R_i F_{id} + m_i^\Delta g c_{3,3} + m_i^\Delta \ddot{p}_i^I\|_\infty \\ &\leq \lambda_{i,\Delta p1} \|u_{ip}\|_\infty + \lambda_{i,\Delta p2}, \end{aligned} \tag{6.26}$$

where $\lambda_{i,\Delta p1}$ and $\lambda_{i,\Delta p2}$ are positive constants. Combining (6.15), (6.16), and (6.14) leads to

$$\|u_{ip}\|_\infty \leq \lambda_{i,up1} \|e_{ip}\|_\infty + \lambda_{i,up2} \|\Delta_{ip}\|_\infty + \lambda_{i,up3}, \tag{6.27}$$

where $\lambda_{i,up1}$, $\lambda_{i,up2}$, and $\lambda_{i,up3}$ are positive constants. Combining (6.26) and (6.27) yields that

$$\|\Delta_{ip}\|_\infty \leq \lambda_{i,\Delta u1} \|e_p\|_\infty + \lambda_{i,\Delta u2}.$$

It follows that

$$\|\Delta_p\|_\infty \leq \lambda_{\Delta u1} \|e_p\|_\infty + \lambda_{\Delta u2}, \tag{6.28}$$

where $\lambda_{\Delta u1} = \max_i \lambda_{i,\Delta u1}$ and $\lambda_{\Delta u2} = \max_i \lambda_{i,\Delta u2}$. One can see that if $f_{pm} \geq f_p^*$, such that $f_{pm} \geq \lambda_{\Delta u1} \lambda_p$, the following inequalities hold

$$\begin{aligned} \|\Delta_p\|_\infty &\leq \frac{\lambda_{\Delta u1}\left(\|\pi_{p0}\|_\infty + \chi_{p0}\chi_0\right) + \lambda_{\Delta u2}}{1 - \lambda_{\Delta u1} \lambda_p f_{pm}^{-1}}, \\ \|e_p\|_\infty &\leq \frac{\|\pi_{p0}\|_\infty + \chi_{p0}\chi_0 + \lambda_{\Delta u2} \lambda_p f_{pm}^{-1}}{1 - \lambda_{\Delta u1} \lambda_p f_{pm}^{-1}}. \end{aligned} \tag{6.29}$$

Since the second derivative $\ddot{p}_0^r$ of the desired trajectory of the virtual leader is bounded and converges to 0 in a finite time and the matrix A_p is asymptotically stable, there exist positive constants $\varepsilon_{\Delta p}$ and ε_{ep} such that

$$\begin{aligned} \left\|\Delta_p\right\|_\infty &\le \varepsilon_{\Delta p}, \\ \left\|e_p\right\|_\infty &\le \varepsilon_{ep} \end{aligned} . \tag{6.30}$$

From (6.25) and (6.30), one can obtain that

$$\max_j \left|c_{6N,j}^T e_p(t)\right| \le \max_j \left|c_{6N,j}^T \pi_{p0}\right| + \chi_{p0}\chi_0 + f_{pm}^{-1}\lambda_p \varepsilon_{\Delta p}. \tag{6.31}$$

In (6.31), for a given positive constant ε_p, there exist a positive constant T_p^* and a positive parameter f_p^*, such that if $f_{i,pj} \ge f_p^*$ $(j = 1,2,3)$, the tracking error is bounded and satisfies that $\max_j \left|e_{i,pj}(t)\right| \le \varepsilon_p, \forall t \ge T_p^*$.

According to (6.13) and (6.19), one can obtain that

$$\begin{aligned} \left\|\Delta_{ia}\right\|_\infty \le& \left\|B_{ia}\right\|_1 \left\|\Lambda_{i3}\Gamma_{i,r1}U_{i1} + \Lambda_{i2}\Gamma_{i,a1}U_{i2}\right\|_\infty \\ &+ \left\|B_{ia}\right\|_1 \left\|\omega_i^\times J_i^\Delta \omega_i\right\|_\infty + \left\|B_{ia}\right\|_1 \left\|J_i^\Delta \dot{\omega}_i\right\|_\infty + \left\|B_{ia}\right\|_1 \left\|M_{id}\right\|_\infty \\ &\le \lambda_{i,\Delta a1}\left\|u_{ia}\right\|_\infty + \lambda_{i,\Delta a2}\left\|e_{ia}\right\|_\infty^2 + \lambda_{i,\Delta a3}\left\|e_{ia}\right\|_\infty + \lambda_{i,\Delta a4}, \end{aligned} \tag{6.32}$$

where $\lambda_{i,\Delta a1}$, $\lambda_{i,\Delta a2}$, and $\lambda_{i,\Delta a3}$ are positive constants.

According to (6.19), (6.20), (6.21), and (6.22), one can have that

$$\begin{aligned} \left\|u_{ia}\right\|_\infty \le& \left\|\omega_i^\times J_i^N \omega_i\right\|_\infty + \left\|B_{ia}^{-1}\right\|_1 \left\|K_q e_{iQ} + K_w e_{iw}\right\|_\infty \\ &+ \left\|B_{ia}^{-1}\right\|_1 \left\|\Delta_{ia}\right\|_\infty + \left\|B_{ia}^{-1}\dot{\omega}_i^r + M_{iw} + M_{ig}\right\|_\infty \\ &\le \lambda_{i,ua1}\left\|e_{ia}\right\|_\infty^2 + \lambda_{i,ua2}\left\|e_{ia}\right\|_\infty + \lambda_{i,ua3}\left\|\Delta_{ia}\right\|_\infty + \lambda_{i,ua4}, \end{aligned} \tag{6.33}$$

where $\lambda_{i,ua1}$, $\lambda_{i,ua2}$, $\lambda_{i,ua3}$, and $\lambda_{i,ua4}$ are positive constants. Combining (6.32) and (6.33) leads to

$$\left\|\Delta_{ia}\right\|_\infty \le \lambda_{i,\Delta e1}\left\|e_{ia}\right\|_\infty^2 + \lambda_{i,\Delta e2}\left\|e_{ia}\right\|_\infty + \lambda_{i,\Delta e3}, \tag{6.34}$$

where $\lambda_{i,\Delta e1}$, $\lambda_{i,\Delta e2}$, and $\lambda_{i,\Delta e3}$ are positive constants. It follows that

$$\left\|\Delta_a\right\|_\infty \le \lambda_{\Delta e1}\left\|e_a\right\|_\infty^2 + \lambda_{\Delta e2}\left\|e_a\right\|_\infty + \lambda_{\Delta e3}, \tag{6.35}$$

where $\lambda_{\Delta e1} = \max_i \lambda_{i,\Delta e1}$, $\lambda_{\Delta e2} = \max_i \lambda_{i,\Delta e2}$, and $\lambda_{\Delta e3} = \max_i \lambda_{i,\Delta e3}$.

If the following inequality holds

$$\lambda_{\Delta e1}\|e_a\|_\infty + \lambda_{\Delta e2} \leq 1/\left(\sqrt{\chi_a} + \chi_a\right), \tag{6.36}$$

one can have the following equation by combining (6.25) and (6.35)

$$\|\Delta_a\|_\infty \leq \pi_{a0}/\sqrt{\chi_a} + \left(1+\sqrt{\chi_a}\right)\lambda_{\Delta e3}. \tag{6.37}$$

From (6.25) and (6.37), it follows that

$$\|e_a\|_\infty \leq \pi_{a0} + \sqrt{\chi_a}\lambda_{ea}, \tag{6.38}$$

where $\lambda_{ea} \geq \pi_{a0} + \sqrt{\chi_a}\left(1+\sqrt{\chi_a}\right)\lambda_{\Delta e3}$. For (6.36), one can obtain the attractive region of $e_a(t)$ as

$$\left\{e_a(t) : \|e_a\|_\infty \leq \lambda_{\Delta e1}^{-1}\left(\chi_a + \sqrt{\chi_a}\right)^{-1} - \lambda_{\Delta e1}^{-1}\lambda_{\Delta e2}\right\}. \tag{6.39}$$

From (6.37) and (6.25), one can have that

$$\max_j\left|c_{6N,j}^T e_a(t)\right| \leq \max_j\left|c_{6N,j}^T \pi_{a0}\right| + \sqrt{f_{pm}^{-1}\lambda_a}\,\lambda_{ea}.$$

From the above analysis, it can be observed that for a given positive constant ε_a and the given initial condition, there exist a positive constant T_a^* and a positive parameter f_a^*, such that if $f_{i,aj} \geq f_a^*$ $(j=1,2,3)$, the tracking error of the translational system is bounded and satisfies that $\max_j\left|e_{i,aj}(t)\right| \leq \varepsilon_a, \forall t \geq T_a^*$.

Remark 6.3

The influence of the multiple faults in actuators is included in the equivalent disturbances in Δ_{ip} and Δ_{ia}, which can be restrained by the constructed fault-tolerant formation controller with no need for any fault knowledge. Furthermore, the influence of nonlinearities, parametric perturbation, and airflow disturbances can be restrained by the designed distributed controller, simultaneously. Moreover, the trajectory and attitude tracking errors of the global closed-loop control system can converge into a given neighborhood of the origin ultimately by selecting proper $f_{i,pj}$ and $f_{i,aj}$ $(j=1,\ 2,\ 3)$.

6.5 Simulation Results

In order to verify the effectiveness of the proposed fault-tolerant formation control method, multiple tail-sitters are simulated under the influence of multiple actuator faults and uncertainties. Four tail-sitters (tail-sitter 1, tail-sitter 2, tail-sitter 3, and tail-sitter 4) are considered in the formation. For the tail-sitter formation system, the following fault modes are considered: Tail-sitter 1 and tail-sitter 3 have no faults, and tail-sitter 2 and tail-sitter 4 suffer from actuator faults at 12 s, simultaneously. For tail-sitter 2, all actuators are subject to a 30% loss of effectiveness. The failure models can be given as

$$\sigma_{2,rk}, \sigma_{2,aj}(t) = \begin{cases} 0, & 0 < t < 12, \\ 0.3, & t \geq 12, \quad k = 1,2,3,4, \ j = 1,2. \end{cases}$$

For tail-sitter 4, Rotor 4 is completely stuck and the two ailerons are subject to a 24% loss of effectiveness in the control torque, respectively. The failure models can be given as

$$\sigma_{4,r4} = \begin{cases} 0, & 0 < t < 12, \\ 1, & t \geq 12, \end{cases}$$

$$\sigma_{4,rj}, \sigma_{4,ak}(t) = \begin{cases} 0, & 0 < t < 12, \\ 0.24, & t \geq 12, \quad j = 1,2,3, k = 1,2. \end{cases}$$

The trajectory of the virtual leader is given as $p_{y0}^r = 0$, and

$$p_{x0}^r = \begin{cases} 0, & t \leq 10, \\ t^2 - 20t + 100, & 10 < t \leq 20, \\ 20t - 300, & t > 20, \end{cases}$$

$$p_{z0}^r = \begin{cases} t, & t \leq 10, \\ 40 - 30e^{-0.125(t-10)^2}, & t > 10. \end{cases}$$

Each real vehicle parameter is assumed to be 25% larger than its nominal value, respectively. The time-varying external disturbances acting on tail-sitter i are given by

$$F_{id} = \begin{bmatrix} 11.2\sin(2t) + 15.8\cos(t) \\ 13\sin(3t) + 21\cos(3t) \\ 5.6\sin(t) + 27.5\cos(2t) \end{bmatrix},$$

$$M_{id} = \begin{bmatrix} 8.5\sin(2t) + 3.6\cos(2t) \\ 10.6\sin(t) + 7.3\cos(3t) \\ 13.4\sin(2t) + 9.7\cos(3t) \end{bmatrix}.$$

The nominal controller parameters are chosen as $K_q = diag\{130,120,120\}$, $K_w = diag\{105.3,100.8,100.8\}$, $K_p = diag\{3.5,3.5,3.5\}$, and $K_v = diag\{1.5,1.5,1.5\}$. The robust compensator parameters are chosen as $f_{i,p1} = 1000$, $f_{i,p2} = 1000$, $f_{i,p3} = 1000$, $f_{i,a1} = 50$, $f_{i,a2} = 50$, and $f_{i,a3} = 50$. The directed communication topology is selected as the node set $V = \{v_1, v_2, v_3, v_4\}$, the edge set $E = \{(v_4, v_3), (v_3, v_2), (v_2, v_1)\}$, and the weighted matrix $W = [w_{ij}]$ with 0–1 weights. In the task, tail-sitter 4 is considered as the root of the directed graph; thereby, $k_{1g} = 1$, $k_{2g} = 0$, $k_{3g} = 0$, and $k_{4g} = 0$. In order to achieve the time-varying formation flight, the desired deviations of the four tail-sitters can be chosen as

$$h_1 = \begin{cases} \begin{bmatrix} 0 & 4 & 0 \end{bmatrix}^T, t < 10, \\ \begin{bmatrix} 0 & 5 - e^{-(t-10)} & 0 \end{bmatrix}^T, t \le 10, \end{cases}$$

$$h_2 = \begin{cases} \begin{bmatrix} -4 & 0 & 0 \end{bmatrix}^T, t < 10, \\ \begin{bmatrix} -4e^{-(t-10)} & 0 & 0 \end{bmatrix}^T, t \le 10, \end{cases}$$

$$h_3 = \begin{cases} \begin{bmatrix} 0 & -4 & 0 \end{bmatrix}^T, t < 10, \\ \begin{bmatrix} 0 & e^{-(t-10)} - 5 & 0 \end{bmatrix}^T, t \le 10, \end{cases}$$

$$h_4 = \begin{cases} \begin{bmatrix} 4 & 0 & 0 \end{bmatrix}^T, t < 10, \\ \begin{bmatrix} 4e^{-(t-10)} & 0 & 10 - 10e^{-(t-10)} \end{bmatrix}^T, t \le 10. \end{cases}$$

The simulation results for the time-varying formation pattern subject to communication delays and multiple uncertainties using the proposed control method are depicted in Figures 6.3–6.5. Figure 6.3 depicts the trajectories of the group tail-sitters. The attitude and position response are shown in Figures 6.4 and 6.5, respectively. Furthermore, a baseline nominal controller developed from [116] is implemented for comparisons. The tracking errors of the position by the two control methods are compared in Figures 6.6 and 6.7, respectively. From these results, one can see that the position tracking errors using the proposed controller are smaller than the baseline nominal controller, and the attitude tracking performance by the proposed controller can be recovered back to normal response quickly when the faults occur. Moreover, the real formation trajectories by the baseline nominal controller deviate

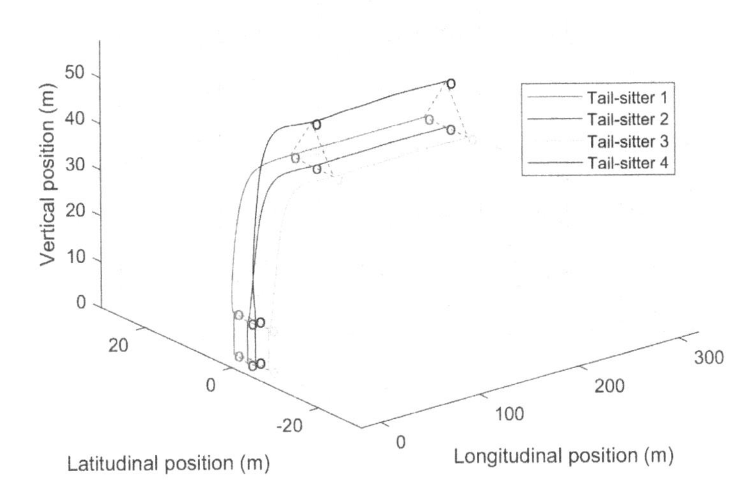

FIGURE 6.3
Formation trajectory using the proposed robust controller.

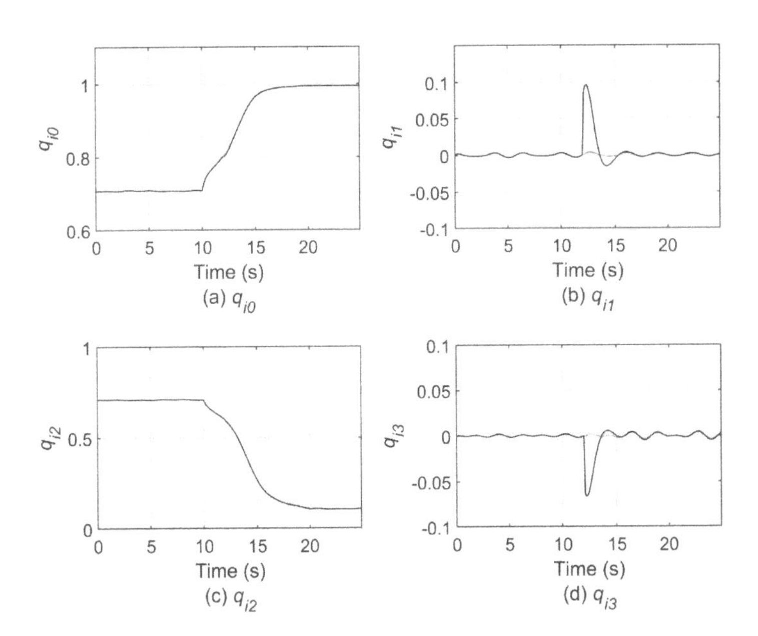

FIGURE 6.4
Attitude response using the proposed robust controller.

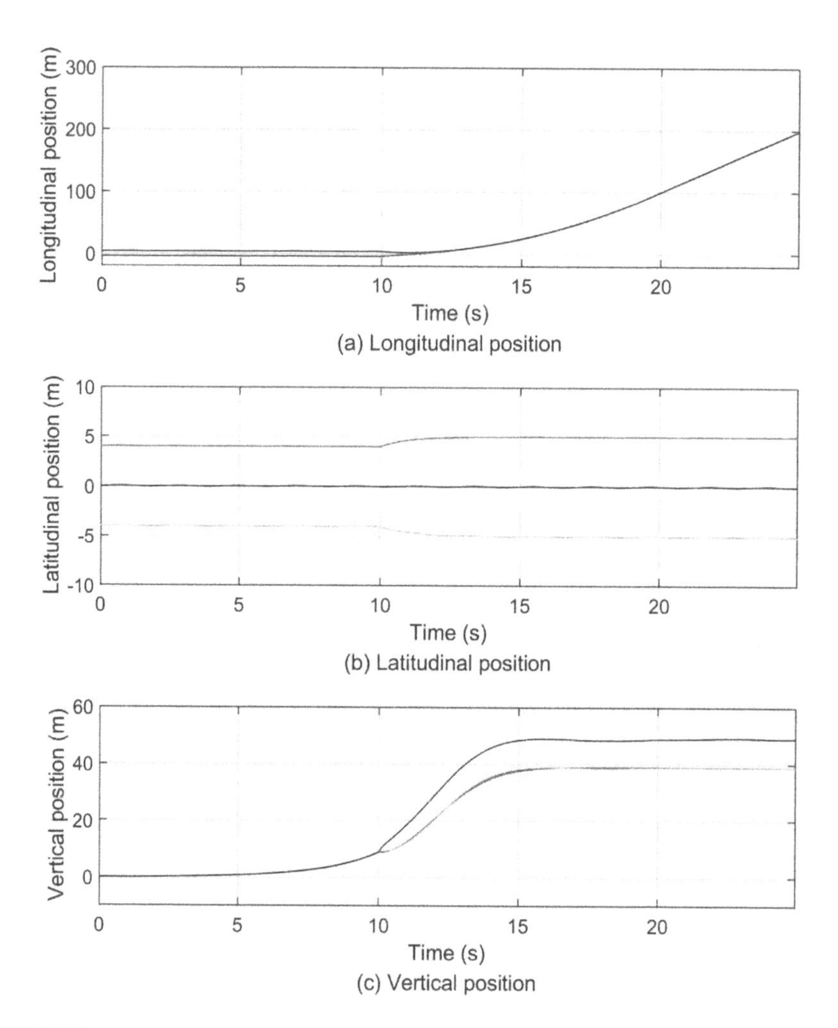

FIGURE 6.5
Position response using the proposed controller.

from the desired trajectories. However, the proposed controller can achieve the predefined time-varying formation scenario. The proposed robust fault-tolerant formation control method can restrain the effects of the multiple actuator faults, parametric perturbations, coupling, nonlinear dynamics, and external disturbances simultaneously.

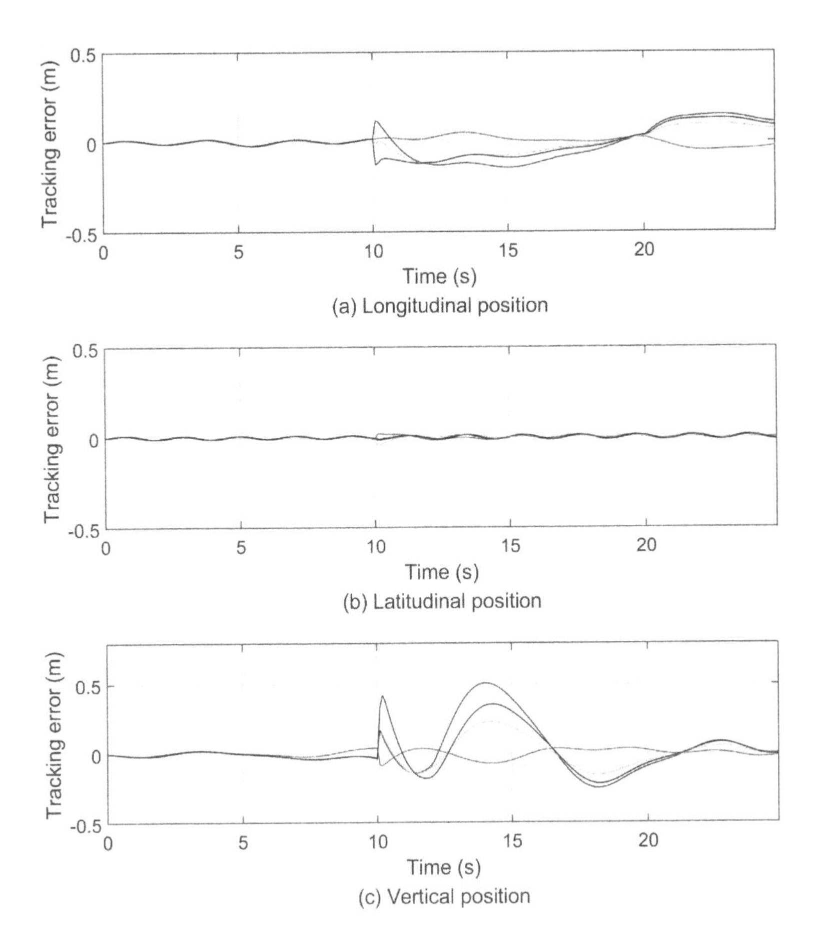

FIGURE 6.6
Trajectory tracking errors using the proposed robust controller.

6.6 Conclusion

In this chapter, a robust distributed fault-tolerant and continuous time-varying formation control approach is proposed for a group of tail-sitters subject to multiple actuator faults and uncertainties in aggressive flight mode transitions. For each tail-sitter, the proposed controller can be divided into an

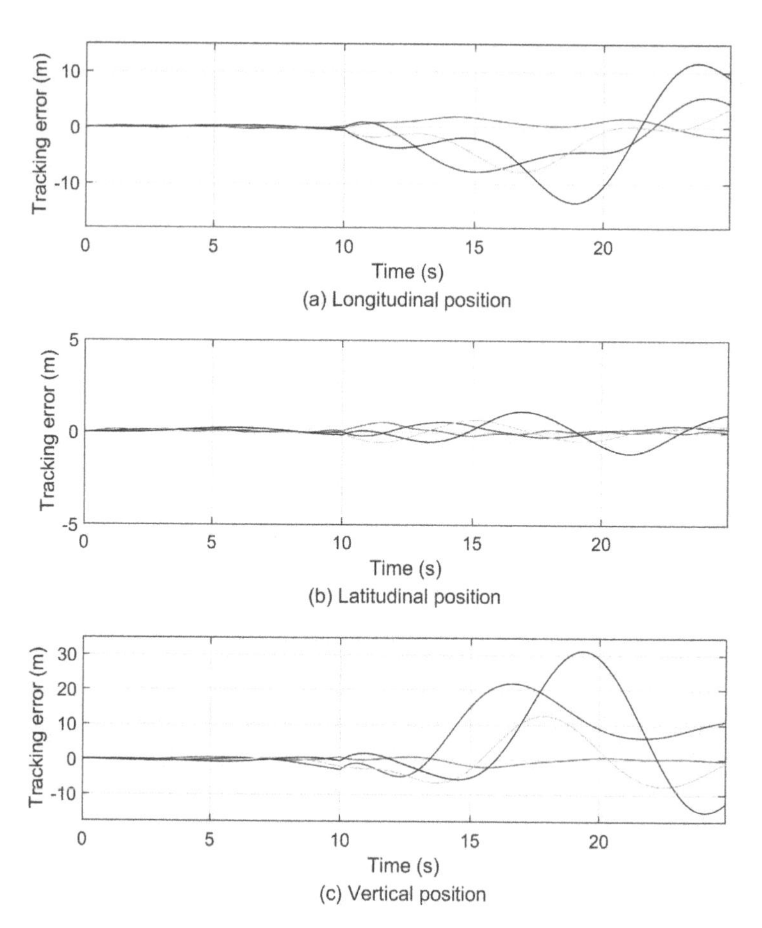

FIGURE 6.7
Trajectory tracking errors using the baseline controller.

inner attitude controller and an outer position controller. The controller can achieve the time-varying formation flight in the flight mode transitions with no switching requirements on the coordinate systems or the controller, and the information of the actuator faults does not need to be identified online. Furthermore, it is proven that the stability of the proposed global closed-loop system can be guaranteed in the presence of multiple actuator faults and uncertainties. The simulation results demonstrate the effectiveness of the proposed fault-tolerant time-varying formation control method.

Bibliography

1. Y. Gu, B. Seanor, G. Campa, M. R. Napolitano, L. Rowe, S. Gururajan and S. Wan. Design and flight testing evaluation of formation control laws. *IEEE Transactions on Control System Technology*, 14(6): 1105–1112, 2006.
2. V. Roldao, R. Cunba, D. Cabecinhas, C. Silvestre, and P. Oliveira. A leader-following trajectory generator with application to quadrotor formation flight. *Robotics and Autonomous Systems*, 62(10): 1597–1609, 2014.
3. K. Yamamoto, K. Sekiguchi, K. Nonaka. Experimental verification of formation control by model predictive control considering collision avoidance in three dimensional space with quadcopters. *11th Asian Control Conference (ASCC)*, 2017.
4. B. Wang, X. Dong, B. M. Chen, T. H. Lee, and S. K. Phang. Formation Flight of unmanned rotorcraft based on robust and perfect tracking approach. *Proceedings of the 2012 American Control Conference*, 3284–3290, 2012.
5. N. Michael, D. Mellinger, Q. Lindse, V. Kumar. The GRASP multiple micro-UAV testbed. *IEEE Robotics and Automation Magazine*, 17(3):56–65, 2010.
6. A. Kushleyev, D. Mellinger, C. Powers, and V. Kumar. Towards a swarm of agile quadrotors. *Autonomous Robots*, 35(4):287–300, 2013.
7. J. Shen, Z. Ning, C. Bo, H. Rui, and L. Zhang. A novel control approach for piecewise-affine systems with quantization in both measurement outputs and control inputs. *2016 Seventh International Conference on Intelligent Control and Information Processing*, 111–116, 2016.
8. S. Hauert, S. Leven, M. Varga, F. Ruini, A. Cangelosi, J.-C. Zufferey, and D. Floreano. Reynolds flocking in reality with fixed-wing robots: Communication range vs. maximum turning rate. *Proceedings of the 2011 IEEE/RSJ International Conference on Intelligent Robots and Systems*, 5015–5020, 2011.
9. S.-M. Kang and H.-S. Ahn. Design and realization of distributed adaptive formation control law for multi-agent systems with moving leader. *IEEE Transactions on Industrial Electronic*, 63(2): 1268–1279, 2016.
10. B. Yu, X. Dong, Z. Shi, and Y. Zhong. Formation control for quadrotor swarm systems: Algorithms and experiments. *Proceedings of the 32nd Chinese Control Conference*, 7099–7104, 2013.
11. X. Dong, Y. Zhou, R. Zhang and Y. Zhong. Time-varying formation tracking for second-order multi-agent systems subjected to switching topologies with application to quadrotor formation flying. *IEEE Transactions on Industrial Electronics*, 64(6): 5014–5024, 2017.
12. Xu Z, He F, Xing X, H. Qi, and X. Huo. Modelling and control of a quadrotor equipped with an unbalanced load. *Asian Control Conference*. 2017:784–789.
13. A. Karimoddini, H. Lin, B. M. Chen, and T. H. Lee. Hybrid three dimensional formation control for unmanned helicopters. *Automatica*, 49(2): 424–433, 2013.
14. A. Mahmood and Y. Kim. Leader-following formation control of quadcopters with heading synchronization. *Aerospace Science and Technology*, 47(1): 68–74, 2015.

15. C. Hua, J. Chen, and Y. Li. Leader-follower finite-time formation control of multiple quadrotors with prescribed performance. *International Journal of Systems Science*, 48(12): 2499–2508, 2017.
16. W. Jasim and D. Gu. Robust team formation control for quadrotors. *IEEE Transactions on Control System Technology*, 26(4): 1516–1523, 2018.
17. M. Mattei and V. Scordamaglia. Task priority approach to the coordinated control of a team of flying vehicles in the presence of obstacles. *IET Control Theory and Applications*, 6(13): 2103–2110, 2012.
18. S. Kim and Y. Kim. Optimum design of three-dimensional behavioral decentralized controller for UAV formation flight. *Engineering Optimization*, 41(3): 199–224, 2009.
19. L. Garcia-Delgado, A. Dzul, V. Santibanez, and M. Llama. Quad-rotors formation based on potential functions with obstacle avoidance. *IET Control Theory and Applications*, 6(12): 1787–1802, 2012.
20. W. Ren. Consensus strategies for cooperative control of vehicle formations. *IET Control Theory and Applications*, 1(2): 505–512, 2007.
21. M. Turpin, N. Michael, and V. Kumar. Trajectory design and control for aggressive formation flight with quadrotors. *Autonomous Robots*, 33(1–2):143–156, 2012.
22. B. Zhu, H. H.-T. Liu, and Z. Li. Robust distributed attitude synchronization of multiple three-DOF experimental helicopters. *Control Engineering Practice*, 36(3): 87–99, 2015.
23. Z. Li, X. Xing, and J. Yu. Decentralized output-feedback formation control of multiple 3-DOF laboratory helicopters. *Journal of The Franklin Institute-Engineering and Applied Mathematics*, 352(9): 3827–3842, 2015.
24. H. Du, W. Zhu, G. Wen, Z. Duan, and J. Lv. Distributed formation control of multiple quadrotor aircraft based on nonsmooth consensus algorithms. *IEEE Transactions on Cybernetics*, 49(1): 342–353, 2019.
25. J. Huang, C. Wen, W. Wang, and Y.-D. Song. Adaptive finite-time consensus control of a group of uncertain nonlinear mechanical systems. *Automatica*, 51(1): 292–301, 2015.
26. Y. Rochefort, H. Piet-Lahanier, S. Bertrand, D. Beauvois, and D. Dumur. Model predictive control of cooperative vehicles using systematic search approach. *Control Engineering Practice*, 32(SI): 204–217, 2014.
27. A. Abdessameud and A. Taybi. Formation control of VTOL unmanned aerial vehicles with communication delays. *Automatica*, 47(11): 2383–2394, 2011.
28. W. Zhu and D. Cheng. Leader-following consensus of second-order agents with multiple time-varying delays. *Automatica*, 46(12): 1994–1999, 2010.
29. W. Ni and D. Z. Cheng. Leader-following consensus of multi-agent systems under fixed and switching topologies. *System & Control Letters*, 59(3–4): 209–217, 2010.
30. J. Seo, Y. Kim, S. Kim, and A. Tsourdos. Consensus-based reconfigurable controller design for unmanned aerial vehicle formation flight. *Aerospace Science and Technology*, 226(7): 817–829, 2012.
31. X. Dong, B. Yu, Z. Shi, and Y. Zhong. Time-varying formation control for unmanned aerial vehicles: theories and applications. *IEEE Transactions on Control Systems Technology*, 23(1): 340–348, 2015.
32. M. A. Kamel, K. A. Ghamry, Y. Zhang. Real-time fault-tolerant cooperative control of multiple UAVs-UGVs in the presence of actuator faults. *Journal of Intelligent & Robotic Systems*, 88(1): 469–480. 2017.

33. M. Saska, T. Krajnik, V. Vonasek, Z. Kasl, V. Spurny, and L. Preucil. Fault-tolerant formation driving mechanism designed for heterogeneous MAVs-UGVs groups. *Journal of Intelligent & Robotic Systems*, 73(1): 603–622, 2014.
34. Z. Yu, Y. Qu, and Y. Zhang. Safe control of trailing UAV in close formation flight against actuator fault and wake vortex effect. *Aerospace Science and Technology*, 77(1): 189–205, 2018.
35. H. Yang, B. Jiang, H. Yang, and K. Zhang. Cooperative control reconfiguration in multiple quadrotor systems with actuator faults. *IFAC-PapersOnLine*, 48(21): 386–391, 2015.
36. D. Richert, and J. Cortes. Optimal leader allocation in UAV formation pairs ensuring cooperation. *Automatica*, 49(11): 3189–3198, 2013.
37. Y. Hua, X. Dong, Q. Li, and Z. Ren. Distributed time-varying formation robust tracking for general linear multiagent systems with parameter uncertainties and external disturbances. *IEEE Transactions on Cybernetics*, 47(8): 1959–1969, 2017.
38. J. Wang, and M. Xin. Integrated optimal formation control of multiple unmanned aerial vehicles. *IEEE Transactions on Control System Technology*, 23(5): 1731–1744, 2013.
39. X. Wang, V. Yadav, and S. N. Balakrishnan. Cooperative UAV formation flying with obstacle/collision avoidance. *IEEE Transactions on Control System Technology*, 15(4): 672–679, 2007.
40. N. Nigam, S. Bieniawski, I. Kroo, and J. Vian. Control of multiple UAVs for persistent surveillance: algorithm and flight test results. *IEEE Transactions on Control System Technology*, 20(5): 1236–1251, 2012.
41. W. R. Williamson, M. F. Abdel-Hafez, I. Rhee, E.-J. Song, J. D. Wolfe, D. F. Chichka, and J. L. Speyer. An instrumentation system applied to formation flight. *IEEE Transactions on Control System Technology*, 15(1): 75–85, 2007.
42. T. Balch, and R. C. Arkin. Behavior-based formation control for multirobot teams. *IEEE Transactions on Automatic Control*, 14(6): 926–939, 1998.
43. M. Radmanesh and M. Kumar. Flight formation of UAVs in presence of moving obstacles using fast-dynamic mixed integer linear programming. *Aerospace Science and Technology*, 50(1): 149–160, 2016.
44. A. Karimoddini, H. Lin, B. M. Chen, T. H. Lee. Hybrid three-dimensional formation control for unmanned helicopters. *Automatica*, 49(2): 424–433, 2013.
45. E. Zhao, T. Chao, S. Wang, and M. Yang. Finite-time formation control for multiple flight vehicles with accurate linearization model. *Aerospace Science and Technology*, 71(1): 90–98, 2017.
46. R. W. Beard, J. Lawton, and F. Y. Hadaegh. A coordination architecture for spacecraft formation control. *IEEE Transactions on Control System Technology*, 9(6): 777–790, 2001.
47. H. Du, W. Zhu, G. Wen, and D. Wu. Finite-time formation control for a group of quadrotor aircraft. *Aerospace Science and Technology*, 69(1): 609–616, 2017.
48. H. Liu, J. Xi, and Y. Zhong. Robust attitude stabilization for nonlinear quadrotor systems with uncertainties and delays. *IEEE Transactions on Industrial Electronics*, 64(7): 5585–5594, 2017.
49. H. Liu, X. Wang, and Y. Zhong. Quaternion-based robust attitude control for uncertain robotic quadrotors. *IEEE Transactions on Industrial Informatics*, 11(2): 406–415, 2015.

50. A. Das, F. L. Lewis, and K. Subbarao. Backstepping approach for controlling a quadrotor using Lagrange form dynamics. *Journal of Intelligent & Robotic Systems*, 56(1–2): 127–151, 2009.
51. H. Liu, D. Li, Z. Zuo, and Y. Zhong. Robust three-loop trajectory tracking control for quadrotors with multiple uncertainties. *IEEE Transactions on Industrial Electronics*, 63(4): 2263–2274, 2016.
52. H. Zhang, F. L. Lewis, and A. Das. Optimal design for synchronization of cooperative systems: state feedback, observer and output feedback. *IEEE Transactions on Automatic Control*, 56(8): 1948–1952, 2011.
53. S. Khoo, L. Xie, and Z. Man. Robust finite-time consensus tracking algorithm for multirobot systems. *IEEE/ASME Transactions on Mechatronics*, 14(2): 219–228, 2009.
54. I. Bayezit and B. Fidan. Distributed cohesive motion control of flight vehicle formations. *IEEE Transactions on Industrial Electronics*, 60(12): 5763–5772, 2013.
55. F. Liao, R. Teo, J. L. Wang, X. Dong, F. Lin, and K. Peng. Distributed formation and reconfiguration control of VTOL UAVs. *IEEE Transactions on Control Systems Technology*, 25(1): 270–277, 2017.
56. H. Liu, T. Ma, F. L. Lewis, and Y. Wan. Robust formation control for multiple quadrotors with nonlinearities and disturbances. *IEEE Transactions on Cybernetics*, 50(4): 1362–1371, 2020.
57. S.-M. Kang, M.-C. Park, and H.-S. Ahn. Distance-based cycle-free persistent formation: global convergence and experimental test with a group of quadcopters. *IEEE Transactions on Industrial Electronics*, 64(1): 380–389, 2017.
58. W. Ren and N. Sorensen. Distributed coordination architecture for multi-robot formation control. *Robotics and Autonomous Systems*, 56(44): 324–333, 2008.
59. B. Xu. Composite learning finite-time control with application to quadrotors. *IEEE Transactions on Systems, Man, and Cybernetics: Systems*, 48(10): 1806–1815, 2018.
60. M. Rehan, A. Jameel, and C. K. Ahn. Distributed consensus control of one-sided Lipschitz nonlinear multiagent systems. *IEEE Transactions on Systems, Man, and Cybernetics: Systems*, 48(8): 1297–1308, 2018.
61. X. Lin and Y. Zheng. Finite-time consensus of switched multiagent systems. *IEEE Transactions on Systems, Man, and Cybernetics: Systems*, 47(7): 1535–1545, 2017.
62. Y. Zhong. Robust output tracking control of SISO plants with multiple operating points and with parametric and unstructured uncertainties. *International Journal of Control*, 75(4): 219–241, 2002.
63. P. Tokekar, J. V. Hook, D. Mulla, and V. Isler. Sensor planning for a symbiotic UAV and UGV system for precision agriculture. *IEEE Transactions on Robotics*, 32(6): 1498–1511, 2016.
64. S. Jeong, O. Simeone, and J. Kang. Mobile edge computing via a UAV-mounted cloudlet: optimization of bit allocation and path planning. *IEEE Transactions on Vehicular Technology*, 67(3): 2049–2063, 2018.
65. M. M. Azari, F. Rosas, K.-C. Chen, and S. Pollin. Ultra reliable UAV communication using altitude and cooperation diversity. *IEEE Transactions on Communications*, 66(1): 330–344, 2018.
66. Q. Wu, Y. Zeng, and R. Zhang. Joint trajectory and communication design for multi-UAV enabled wireless networks. *IEEE Transactions on Wireless Communications*, 17(3): 2109–2121, 2018.

67. N. Zhao, F. Cheng, F. R. Yu, J. Tang, Y. Chen, G. Gui, and H. Sari. Caching UAV assisted secure transmission in hyper-dense networks based on interference alignment. *IEEE Transactions on Communications*, 66(5): 2281–2294, 2018.
68. J. Qin, W. Zheng, and H. Gao. Consensus of multiple second-order vehicles with a time-varying reference signal under directed topology. *Automatica*, 47(9): 1983–1991, 2011.
69. S. Yin, H. Yang, and O. Kaynak. Coordination task triggered formation control algorithm for multiple marine vessels. *IEEE Transactions on Industrial Electronics*, 64(6): 4984–4993, 2017.
70. Z. Han, K. Guo, and Z. Lin. Integrated relative localization and leader-follower formation control. *IEEE Transactions on Automatic Control*, 64(1): 20–34, 2019.
71. T. Nguyen, H. M. La, T. D. Le, and M. Jafari. Formation control and obstacle avoidance of multiple rectangular agents with limited communication ranges. *IEEE Transactions on Control of Network Systems*, 4(4): 680–691, 2017.
72. H. Hong, W. Yu, J. Fu, and X. Yu. Finite-time connectivity-preserving consensus for second-order nonlinear multi-agent systems. *IEEE Transactions on Control of Network Systems*, 6(1): 236–248, 2019.
73. A. Loria, J. Dasdemir, and N. A. Jarquin. Leader-follower formation and tracking control of mobile robots along straight paths. *IEEE Transactions on Control Systems Technology*, 24(2): 727–732, 2016.
74. W. Wang, J. Huang, C. Wen, and H. Fan. Distributed adaptive control for consensus tracking with application to formation control of nonholonomic mobile robots. *Automatica*, 50(4): 1254–1263, 2014.
75. S. Li, J. Zhang, F. Wang, and X. Guan. Formation control of heterogeneous discrete-time nonlinear multi-agent systems with uncertainties. *IEEE Transactions on Industrial Electronics*, 64(6): 4730–4740, 2017.
76. B. Liu, T. Chu, L. Wang, and G. Xie. Controllability of a leader-follower dynamic network with switching topology. *IEEE Transactions on Automatic Control*, 53(4): 1009–1013, 2008.
77. X. Dong and G. Hu. Time-varying formation control for general linear multi-agent systems with switching directed topologies. *Automatica*, 73: 47–55, 2016.
78. H. Liu, Y. Tian, F. L. Lewis, Y. Wan, and K. P. Valavanis. Robust formation tracking control for multiple quadrotors under aggressive maneuvers. *Automatica*, 105: 179–185, 2019.
79. W. Ren and R. W. Beard, *Distributed Consensus in Multi-Vehicle Cooperative Control*, London: Springer, 2008.
80. C. Liu, B. Jiang, and K. Zhang. Adaptive fault-tolerant H-infinity output feedback control for lead-wing close formation flight. *IEEE Transactions on Systems, Man, and Cybernetics: Systems*, 50(8): 2804–2814, 2020.
81. Y. Zhao, Q. Duan, G. Wen, D. Zhang, and B. Wang. Time-varying formation for general linear multiagent systems over directed topologies: A fully distributed adaptive technique. *IEEE Transactions on Systems, Man, and Cybernetics: Systems*, 51(1): 532–541, 2021.
82. Y. Liu, J. M. Montenbruck, D. Zelazo, M. Odelga, S. Rajappa, H. H. Bulthoff, F. Allgower, and A. Zell. A distributed control approach to formation balancing and maneuvering of multiple multirotor UAVs. *IEEE Transactions on Robotics*, 34(4): 870–882, 2018.

83. A. Saeed, A. Younes, S. Islam, J. Dias, L. Seneviratne, and G. Cai. A review on the platform design, dynamic modeling and control of hybrid UAVs. *Proceedings of International Conference on Unmanned Aircraft Systems*, 806–815, 2015.
84. X. Wang, Z. Chen, and Z. Yuan. Modeling and control of an agile tail-sitter aircraft. *Journal of the Franklin Institute*, 352(12): 5437–5472, 2015.
85. A. Oosedo, S. Abiko, A. Konno, and M. Uchiyama. Optimal transition from hovering to level-flight of a quadrotor tail-sitter UAV. *Autonomous Robots*, 41(5): 1143–1159, 2017.
86. V. Hrishikeshavan, C. Bogdanowicz, and I. Chopra. Design performance and testing of a quad rotor biplane air vehicle for multi role missions. *International Journal of Air Vehicles*, 6(3): 155–173, 2014.
87. D. Kubo and S. Suzuki. Tail-sitter vertical takeoff and landing unmanned aerial vehicle: transitional flight analysis. *Journal of Aircraft*, 45(1): 292–297, 2008.
88. Y. Jung and D. H. Shim. Development and application of controller for transition flight of tail-sitter UAV. *Journal of Intelligent & Robotic Systems*, 65(1): 137–152, 2012.
89. M. Hochstenbach, C. Notteboom, B. Theys, and J. D. Schutter. Design and control of an unmanned aerial vehicle for autonomous parcel delivery with transition from vertical take-off to forward flight-VertiKUL, a quadcopter tailsitter. *International Journal of Air Vehicles*, 7(4): 395–406, 2015.
90. J. Zhou, X. Lyu, Z. Li, S. Shen, and F. Zhang. A unified control method for quadrotor tail-sitter UAVs in all flight modes: hover, transition, and level flight. *Proceedings of IEEE/RSJ International Conference on Intelligent Robots and Systems*, 4835–4841, 2017.
91. J. L. Forshaw and V. J. Lappas. Transitional control architecture and methodology for a twin rotor tailsitter. *Journal of Guidance, Control, and Dynamics*, 37(4): 1289–1298, 2014.
92. S. Swarnkar, H. Parwana, M. Kothari, and A. Abhishek. Biplane-quadrotor tail-sitter UAV: flight dynamics and control. *Journal of Guidance, Control, and Dynamics*, 41(5): 1049–1067, 2018.
93. R. Stone, P. Anderson, C. Hutchison, A. Tsai, P. Gibbens, and K. Wong. Flight testing of the T-wing tail-sitter unmanned air vehicle. *Journal of Aircraft*, 45(2), 673–685, 2008.
94. K. Wang, Y. Ke, and B. M. Chen. Autonomous reconfigurable hybrid tail-sitter UAV U-Lion. *Science China (Information Sciences)*, 60(3): 1–16, 2017.
95. A. Banazadeh and N. Taymourtash. Optimal control of an aerial tail sitter in transition flight phases. *Journal of Aircraft*, 53(4): 914–921, 2016.
96. H. Liu, F. Peng, F. L., Lewis, and Y. Wan. Robust tracking control for tail-sitters in flight mode transitions. *IEEE Transactions on Aerospace and Electronic Systems*, 55(4): 2023–2035, 2019.
97. B. Wang, J. Wang, B. Zhang, and X. Li. Global cooperative control framework for multiagent systems subject to actuator saturation with industrial applications. *IEEE Transactions on Systems, Man, and Cybernetics: Systems*, 47(7): 1270–1283, 2017.
98. R. Wang, X. Dong, Q. Li, and Z. Ren. Distributed time-varying formation control for linear swarm systems with switching topologies using an adaptive output-feedback approach. *IEEE Transactions on Systems, Man, and Cybernetics: Systems*, 49(12): 2664–2675, 2012.

99. W. Zhao and T. H. Go. Quadcopter formation flight control combining MPC and robust feedback linearization. *Journal of the Franklin Institute*, 351(3): 1335–1355, 2014.
100. Z. Hou and I. Fantoni. Interactive leader–follower consensus of multiple quadrotors based on composite nonlinear feedback control. *IEEE Transactions on Control Systems Technology*, 26(5): 1732–1743, 2018.
101. H. Liu, Y. Wang, and F. L. Lewis. Robust distributed formation controller design for a group of unmanned underwater vehicles. *IEEE Transactions on Systems, Man, and Cybernetics: Systems*, 51(2): 1215–1223, 2021.
102. S. He, M. Wang, S. Dai, and F. Luo. Leader-follower formation control of USVs with prescribed performance and collision avoidance. *IEEE Transactions on Industrial Informatics*, PP(99): 1–10, 2018.
103. D. J. Bennet, C. R. McInnes, M. Suzuki, and K. Uchiyama. Autonomous three-dimensional formation flight for a swarm of unmanned aerial vehicles. *Automatica*, 34(6): 1899–1908, 2011.
104. Y. Cao, W. Yu, W. Ren, and G. Chen. An overview of recent progress in the study of distributed multi-agent coordination. *IEEE Transactions on Industrial Informatics*, 9(1): 427–438, 2013.
105. J. Rife. Collaborative positioning for formation flight of cargo aircraft. *Journal of Guidance, Control, and Dynamics*, 36(1): 304–307, 2013.
106. L. Qin, X. He, R. Yan, R. Deng, and D. Zhou. Distributed sensor fault diagnosis for a formation of multi-vehicle systems. *Journal of the Franklin Institute*, 356(2): 791–818, *2019*.
107. X. Jin. Fault tolerant finite-time leader-follower formation control for autonomous surface vessels with LOS range and angle constraints. *Automatica*, 68(1): 228–236, 2016.
108. Y. H. Chang, C. I. Wu, and H. W. Lin. Adaptive distributed fault-tolerant formation control for multi-robot systems under partial loss of actuator effectiveness. *International Journal of Control, Automation and Systems*, 16(5): 2114–2124, 2018.
109. S. M. Azizi and K. Khorasani. Cooperative actuator fault accommodation in formation flight of unmanned vehicles using absolute measurements. *IET Control Theory and Applications*, 6(18): 2805–2819, 2011.
110. C. H. Xie and G. H. Yang. Decentralized adaptive fault-tolerant control for large-scale systems with external disturbances and actuator faults. *Automatica*, 85(1): 83–90, 2017.
111. C. Zhang, J. Wang, D. Zhang, and X. Shao. Fault-tolerant adaptive finite-time attitude synchronization and tracking control for multi-spacecraft formation. *Aerospace Science and Technology*, 73(1): 197–209, 2018.
112. A. Oosedo, A. Konno, T. Matsumoto, K. Go, K. Masuko, and M. Uchiyama. Design and attitude control of a quad-rotor tail-sitter vertical takeoff and landing unmanned aerial vehicle. *Advanced Robotics*, 26(3): 307–326, 2012.
113. O. Garcia, P. Castillo, K. C. Wong, and R. Lozano. Attitude stabilization with real-time experiments of a tail-sitter aircraft in horizontal flight. *Journal of Intelligent & Robotic Systems*, 65(1): 123–136, 2012.
114. X. Zhao, Q. Zong, B. Tian, D. Wang, and M. You. Finite-time fault-tolerant formation control for multiquadrotor systems with actuator fault. *International Journal of Robust and Nonlinear Control*, 28(17): 5386–5405, 2018.

115. J. Shi, D. Zhou, Y. Yang, and J. Sun. Fault tolerant multivehicle formation control framework with applications in multiquadrotor systems. *Science China (Information Sciences)*, 61(12): 1–3, 2018.
116. Q. Xu, H. Yang, B. Jiang, D. Zhou, and Y. Zhang. Fault tolerant formations control of uavs subject to permanent and intermittent faults. *Journal of Intelligent & Robotic Systems*, 73(1): 589–602, 2014.
117. Z. Yu, Y. Qu, and Y. Zhang. Fault-tolerant containment control of multiple unmanned aerial vehicles based on distributed sliding-mode observer. *Journal of Intelligent & Robotic Systems*, 93:163–177, 2019.
118. X. Yu, Z. Liu, and Y. Zhang. Fault-tolerant formation control of multiple UAVs in the presence of actuator faults. *International Journal of Robust and Nonlinear Control*, 26(12): 2668–2685, 2015.
119. H. Yang, B. Jiang, H. Yang, and H. H. Liu. Synchronization of multiple 3-DOF helicopters under actuator faults and saturations with prescribed performance. *ISA Transactions*, 75(1): 118–126, 2018.
120. X. Li, M. Z. Q. Chen, H. Su, and C. Li. Distributed bounds on the algebraic connectivity of graphs with application to agent networks. *IEEE Transactions on Cybernetics*, 47(8): 2121–2131, 2017.

Index